Holocene Geomorphic Development of Coastal Ridges in Japan

Akiko Matsubara

KEIO UNIVERSITY PRESS

Holocene Geomorphic Development of Coastal Ridges in Japan

Published by Keio University Press Inc.
19-30, 2-chome, Mita, Minato-ku
Tokyo 108-8346, Japan
Copyright © 2015 by Akiko Matsubara
All rights reserved.

No part of this book may be reproduced in
any manner without the written consent of
the publisher, except in the case of brief
excerpts in critical articles and reviews.

Printed in Japan
ISBN 978-4-7664-2215-3

First edition, 2015

1. Lake Hamana (Chapter 2.1.1)

2. Kujukurihama Lowland (Chapter 2.3.3)

3. Excavation using a Geoslicer in Hamamatsu Lowland
(Chapter 2.1.1)

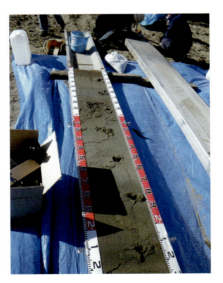

4. Deposits of coastal ridge in Hamamatsu Lowland
(Chapter 2.1.1)

5. Amanohashidate from the southern mountains
(Chapter 2.1.2, 3.3.3)

6. Amanohashidate from the northern mountains
(Chapter 2.1.2, 3.3.3)

**7. Ukishimagahara Lowland at the southern foot of Mt. Fuji
(Chapter 2.2.1)**

**8. Present-day marsh in Ukishimagahara Lowland
(Chapter 2.2.1, 3.1.2)**

9. Sandy deposits of buried coastal ridge in Ukishimagahara Lowland
(Chapter 2.2.1, 3.1.2)

10. Archaeological site on buried coastal ridge in Ukishimagahara Lowland
(Chapter 3.1.2)

11. Shingle beach of Numazu Coast (Chapter 3.3.1)

12. Miho Spit (Chapter 2.3.1)

13. Offshore breakwaters of Miho Spit (Chapter 3.3.2)

14. Offshore breakwaters of Miho Coast (Chapter 3.3.2)

Preface

Coastal ridges such as coastal barriers and beach ridges are widely distributed along stable trailing edge coasts across the globe. The primary requisites for the formation of coastal ridges are sufficient sediment supply, processes allowing the development and maintenance of the ridges, and an appropriate geomorphic setting. Although the Japanese Islands belong to tectonically active areas including island arcs and trench systems, coastal ridge landforms are generally distributed in the coastal lowlands of this area.

Considering the geomorphic development of coastal ridges, studies on barrier complexes worldwide have shown that the rise in sea level during the Holocene was a major factor affecting the development of coastal barriers. Generally, barriers developed and transgressed landward when the sea level was rapidly rising. In contrast, the barriers began to grow seaward when sedimentation rate exceeded the rate of sea level rise. Following this, beach ridges also began to develop in a seaward direction.

A common trend in the relative sea level change around Japan is that the sea rose above the present level as a result of hydroisostatic movements. The Holocene transgression is known as the Jomon transgression in Japan, as it is associated with the archaeological age of that name. During this period, the relative sea level generally reached its maximum (3 to 5 m higher than at present) around 7,000 cal BP. As such, the geomorphic evolution of the coastal lowlands was deeply influenced by these Holocene sea level changes.

First, this study elucidates the geomorphic development of coastal ridges during the Holocene by reconstructing palaeoenvironmental changes on the basis of fossil foraminiferal assemblages analysis. Then, both the common and different processes of coastal ridge development will be discussed. In particular, the influence of sea level changes and tectonic movements on coastal ridge development will be examined.

Second, this study identifies the relationship between coastal ridge development and human activities on the basis of an analysis of the distribution of archaeological and historic sites on the coastal ridges. Furthermore, beach erosion in coastal ridges and the influence of tsunamis on the ridges will be discussed.

Publication of this book was assisted by a grant from Keio Gijuku Fukuzawa Memorial Fund for the Advancement of Education and Reserch.

Akiko Matsubara
December 2014

Contents

Preface i

Introduction 1

Chapter 1 Study Area and Methods of Determining Ridge Development 5
    1.1 Classification and Distribution of Coastal Ridges in Japan 5
    1.2 Reconstruction of Palaeoenvironmental Changes in Coastal Ridges using Fossil Foraminiferal Assemblages 11

Chapter 2 Geomorphic Development of Coastal Ridges in Japan 17
    2.1 Barrier–Lagoon Complexes 17
        2.1.1 Lake Hamana and Hamamatsu Lowland 17
        2.1.2 Amanohashidate 27
    2.2 Sand and Gravel Ridge–Backmarsh Complexes 29
        2.2.1 Kano River and Ukishimagahara Lowlands 29
        2.2.2 Tokoro Lowland 40
        2.2.3 Kuninaka Lowland 44
        2.2.4 Sendai Lowland 51
    2.3 Beach Ridge Plains 52
        2.3.1 Shimizu Lowland 52
        2.3.2 Tateyama Lowland 65
        2.3.3 Kujukurihama Lowland 73
    2.4 Valley Plains 77
        2.4.1 Matsuzaki Lowland 77
        2.4.2 Haibara Lowland 84
    2.5 Delta–Beach Ridge Complexes 92
        2.5.1 Sagami River Lowland 92
        2.5.2 Obitsu River Lowland 97

Chapter 3	Relationship between the Geomorphic Development of Coastal Ridges and Human Activities	**101**
3.1	Distribution of Archaeological Sites on Coastal Ridges	101
	3.1.1 Lake Hamana and Hamamatsu Lowland	101
	3.1.2 Kano River and Ukishimagahara Lowlands	104
	3.1.3 Shimizu Lowland	109
	3.1.4 Haibara Lowland	112
	3.1.5 Other Areas: Tateyama, Kujukurihama, Kuninaka and Obitsu River Lowlands	112
3.2	Distribution of Coastal Ridges and Human Activities in Tokyo	114
	3.2.1 Geomorphic Development of the Kanto Plain	114
	3.2.2 Landforms of Central Areas in Tokyo	116
	3.2.3 Distribution of Archaeological Sites in the Centre of Tokyo	121
3.3	Beach Erosion in Coastal Ridges	123
	3.3.1 Inner Part of Suruga Bay: the Numazu-Fuji Coast	123
	3.3.2 Miho Spit	126
	3.3.3 Amanohashidate	129
3.4	Influence of Tsunamis on Coastal Ridges: Case of the 2011 Tohoku Earthquake	132
Chapter 4	**Geomorphic Development of Coastal Ridges during the Holocene**	**139**
4.1	Common and Different Processes of Coastal Ridge Development	139
	4.1.1 Common Processes in Relation to Relative Sea Level Change	139
	4.1.2 Factors Controlling Different Processes of Ridge Development 1: Tectonic Movements	145
	4.1.3 Factors Controlling Different Processes of Ridge Development 2: Basal Landforms and Sediment Supply	146
4.2	Human Settlement on Coastal Ridges during the Holocene	147

References   151
Index   167

Introduction

The Japanese Islands are composed of arcs and trench systems and are characterized as having been tectonically active during the Quaternary (Yoshikawa et al., 1981). The coastal lowlands of the Japanese Islands are distributed mainly in the subsiding regions. The rivers supply a large amount of sediments to the coastal lowlands, derived from volcanoes or uplifting mountains in their upper reaches. This deposition has been ongoing since the Last Glacial stage in the late Pleistocene, resulting in the accumulation of thick unconsolidated deposits.

The coastal lowlands of the Japanese Islands can be classified into three types: (1) alluvial fans, which develop at river mouths with a large supply of coarse sediments and face the steep sea bottom; (2) alluvial deltas, which dominantly occur in the inner parts of bays, and are supplied with a large amount of fine sediments; and (3) sand and gravel ridge–backmarsh complexes, in which the ridges are usually parallel to the shore and represent former coastal barriers and beach ridges.

Among these three types of coastal lowlands, the ridge–backmarsh complexes are the most extensively distributed. Consequently, it is considered that these coastal ridges have played an important role in the geomorphic evolution of coastal lowlands during the Holocene period.

Studies into barrier systems around the globe have shown that the rise in sea level during the Holocene was a common and major factor driving the development of coastal barriers (Davis, 1994a; Trenhaile, 1997; Sherman, 2013). While the sea level rose rapidly, no stable shoreline persisted long enough to allow the development of barrier islands. However, when the rate of sea level rise slowed, allowing shoreline positions to become more stable and development of coasts. Generally, coastal barriers developed and transgressed landward while the sea level was rising rapidly, whereas the beach ridges began to develop seaward when the sedimentation rate exceeded the rate of sea level rise.

These geomorphic developments of coastal barriers and beach ridges have been recognized along the eastern coast of North America (Colquhoun *et al.*, 1968; Pierce and Colquhoun, 1970; Moslow and Colquhoun, 1981; Moslow and Herson, 1994), the Gulf of Mexico (Wilkinson, 1975; Morton, 1994), the eastern coast of South America (Dillenburg and Hesp, 2009), the North Sea coast of the Netherlands, Germany and Denmark (Van Straaten, 1965; Hageman, 1969; Jelgersma and Van Regteren, 1969; Beets *et al.*, 1992; Davis, 1994b), and the Australian coast (Thom *et al.*, 1981; Thom, 1983; Thom and Roy, 1985; Hesp and Short, 1999; Short, 2010). Beach ridge plains are well developed along the north coast of Canada, the Gulf of Mexico, and the southeast coast of Australia. The development of beach ridges is considered to be related to various physical conditions such as waves, storms, sediment supply, sea level changes, and isostatic crustal movements (Bird and Jones, 1988; Hesp, 1988; Tanner, 1988; Mason, 1993; Mason and Jordan, 1993 ; Taylor and Stone, 1996 ; Sanderson *et al.*, 1998).

A common trend in the relative sea level change around Japan is that the sea rose above the present level (Ota *et al.*, 1981, 1982, 1990; Pirazzoli,

Introduction

1996; Bird, 2008). The lowest stand of the sea level during the Last Glacial period was determined to be between −140 and −80 m around 20,000 cal BP (Kaizuka *et al.*, 1977 and others).

Following this low level, post-glacial sea level rise generally continued until around 7,000 cal BP, with a short-lived regression phase between 12,000 and 11,000 cal BP. It is known that sea level generally reached its highest level, 3 to 5 m higher than at present, around 7,000 cal BP. After the culmination of the Holocene transgression, known locally as the Jomon transgression, the sea level stabilized, or lowered slightly, and has since varied with minor fluctuations (Ota *et al.*, 1981, 1982, 1990; Umitsu, 1991).

The process of coastal lowland evolution comprises three stages, related to relative sea level changes.

(1) The valley-forming stage: Ancestral rivers eroded downwards to form a valley to the depth of the lowest sea level during the latest Pleistocene period.

(2) The bay-forming stage: The valley was invaded by transgressing water from the rise in sea level during the Holocene, and a bay or a drowned valley was formed during the early to mid-Holocene.

(3) The lowland-forming stage: The bay or the drowned valley was filled with fluvial deposits or was enclosed by coastal ridges after the end of the Holocene sea level rise during the late Holocene.

Studies on the geomorphic development of coastal lowlands have mainly dealt with the land-forming stage during the late Holocene. As yet, there is a paucity of studies considering the palaeoenvironmental changes in the bays or drowned valleys, or the development of the coastal ridges. Studies on the coastal ridges in Japan have mainly focused on the development of beach ridges after the sea level had stabilized or lowered slightly during the late Holocene. A small number of studies have addressed the formation of barrier complexes during the sea level rises before the culmination of the transgression in the early to mid-Holocene (Moriwaki, 1979; Suzuki and Saito, 1987). However, these have not sufficiently explained the development of barriers and beach ridges throughout the Holocene. Matsubara (1988, 2000a, 2002, 2005) reconstructed the palaeoenvironmental changes occurring within the embayments along Suruga Bay, on the basis of an analysis of fossil foraminiferal assemblages in bore-hole cores, and used these to describe the geomorphic development of the coastal ridges in this area during the Holocene. These case studies served to indicate the processes involved in the development of coastal barriers while the sea level was rising during the early to mid-Holocene, in addition to those involved in the development of beach ridges when the sea level stabilized or slightly lowered during the late Holocene.

From the perspective of human activities in coastal lowlands, archaeological sites dating from the later Jomon period, around 3,000 BP, are widely distributed on the coastal ridges (Matsubara, 2003). This suggests that the coastal ridges in the lowlands represented important landforms for human settlement during the Holocene.

Chapter 1

Study Area and Methods of Determining Ridge Development

1.1. Classification and Distribution of Coastal Ridges in Japan

The coastal lowlands comprising ridge–backmarsh complexes and coastal lagoons in Japan may be further classified into six types, according to their present-day landforms: (a) barrier–lagoon complexes (barrier systems); (b) sand and gravel ridge–backmarsh complexes; (c) beach ridge plains; (d) valley plains, which may be (d-1) barrier-backmarsh complexes or (d-2) beach ridge plain types; (e) delta-beach ridge complexes; and (f) others, including (f-1) barrier spits and (f-2) cuspate forelands (Fig. 1.1.1, Tab. 1.1.1).

The distribution of each type may be summarized as follows.

(a) Barrier–lagoon complexes (barrier systems): Lagoons develop behind coastal barriers, such as in Lake Saroma (No. 2 in Fig. 1.1.1, Tab. 1.1.1),

CHAPTER 1

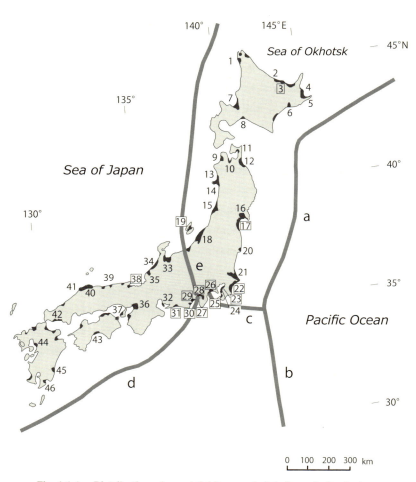

Fig. 1.1.1. Distribution of coastal ridges and plate boundaries in Japan

Numbers in ☐ represent areas studied.
1. Sarobetsu Lowland 2. Lake Saroma 3. Tokoro Lowland 4. Notsuke Spit 5. Lake Furen 6. Kushiro Lowland 7. Ishikari Lowland 8. Yufutsu Lowland 9. Lake Tsugaru-jusan 10. Aomori Lowland 11. Tanabu Lowland 12. Lake Ogawara 13. Noshiro Lowland 14. Akita Lowland 15. Shonai Lowland 16. Ishinomaki Lowland 17. Sendai Lowland 18. Niigata Lowland 19. Lake Kamo and Kuninaka Lowland 20. Iwaki Lowland 21. Lake Kasumigaura 22. Kujukurihama Lowland 23. Obitsu River Lowland 24. Futtsu Lowland 25. Tateyama Lowland 26. Sagami River Lowland 27. Matsuzaki Lowland 28. Kano River and Ukishimagahara Lowlands 29. Shimizu Lowland and Miho Spit 30. Haibara Lowland 31. Lake Hamana and Hamamatsu Lowland 32. Utsumi Lowland 33. Toyama Lowland 34. Lake Kahokugata 35. Fukui Lowland 36. Osaka and Kawachi Lowlands 37. Mihara Lowland 38. Amanohashidate 39. Tottori Lowland 40. Lake Nakaumi 41. Lake Shinji and Izumo Lowland 42. Suonada Lowlands 43. Kochi Lowland 44. Tamana Lowland 45. Miyazaki Lowland 46. Kimotsuki Lowland
a. Japan Trench b. Izu-Ogasawara Trench c. Sagami Trough d. Suruga-Nankai Trough e. Fossa Magna

Lake Kasumigaura (No. 21), Lake Hamana (No. 31) and Lake Nakaumi (No. 40).

(b) Sand and gravel ridge–backmarsh complexes: Marshes or swamps develop behind coastal ridges, whose widths are narrow. For instance, this type of lowland is developed in the Tokoro (No. 3), Ishikari (No. 7), Niigata (No. 18) and Ukishimagahara (No. 28) areas.

(c) Beach ridge plains: Beach ridges are widely developed, such as in the Kujukurihama (No. 22), Shimizu (No. 29) and Miyazaki (No. 45) lowlands.

(d) Valley plains: Coastal ridges develop to enclose previously buried bays. The (d-1) type (barrier–backmarsh complexes) consists of valley plains with ridge–backmarsh complexes, such as in the Matsuzaki (No. 27) and Suonada (No. 42) lowlands. The (d-2) type (beach ridge plains) comprises valley plains with beach ridges, such as in the Tanabu (No. 11) and Haibara (No. 30) lowlands.

(e) Delta-beach ridge complexes: Both deltas and beach ridges are well developed, such as in the Shonai (No. 15), Obitsu River (No. 23), Sagami River (No. 26) and Izumo (No. 41) lowlands.

(f) Others: The (f-1) type comprises barrier spits such as the Notsuke (No. 4) and Miho (No. 29) spits. The (f-2) type comprises cuspate forelands such as the Futtsu lowland (No. 24).

Here, twelve coastal lowlands and two coastal lagoons, representing typical landforms of the coastal lowlands with ridges in Japan, are investigated (Fig. 1.1.1). Among these, Lake Hamana (No. 31) and Amanohashidate (Lake Aso) (No. 38) belong to barrier–lagoon complexes; the Kano River and Ukishimagahara (No. 28), Tokoro (No. 3), Kuninaka (No. 19), and Sendai (No. 17) lowlands are sand and gravel ridge–backmarsh complexes. The Shimizu (No. 29), Tateyama (No. 25), Hamamatsu (No. 31), and Kujukurihama (No. 22) lowlands belong to the beach ridge plain type, and the Matsuzaki (No. 27) and Haibara (No. 30) lowlands are valley plains. The Sagami River (No. 26) and Obitsu River (No. 23) lowlands comprising

Tab. 1.1.1. Studies on development of coastal ridges in Japan

No.	Lowland or Lake	Type	A	B	Literature
1	Sarobetsu Lowland	b	○	○	Akamatsu et al. (1981), Sakaguchi et al. (1985), Ohira (1995)
2	Lake Saroma	a		○	Hirai (1987, 1994)
3	Tokoro Lowland	b	○		Sakaguchi et al. (1985), Hamano et al. (1985), Matsubara (2000a)
4	Notsuke Spit	f–1			Takano (1978)
5	Lake Furen	a			Ohira et al. (1994)
6	Kushiro Lowland	b			Okazaki (1960 a,b)
7	Ishikari Lowland	b	○		Uesugi・Endo (1973), Matsushita (1979), Sagayama et al. (2010), Ishii et al. (2014)
8	Yufutsu Lowland	c	△		Moriwaki (1982)
9	Lake Tsugaru-jusan	a		○	Onuki et al. (1963), Minoura・Nakaya (1990), Koiwa et al. (2014), Kasai・Koiwa (2014)
10	Aomori Lowland	b			Matsumoto (1984)
11	Tanabu Lowland	d–2			Matsumoto (1984)
12	Lake Ogawara	a		○	Hirai (1983, 1994), Ishizuka・Kashima (1986)
13	Noshiro Lowland	a,b		○	Mii (1960), Shiraishi (1990)
14	Akita Lowland	c		○	Mii (1960), Moriwaki (1982), Matsumoto (1984), Shiraishi (1990)
15	Shonai Lowland	e	○		Ariga (1984)
16	Ishinomaki Lowland	c	○		Matsumoto (1984), Ito (1999, 2003)
17	Sendai Lowland	b			Matsumoto (1981, 1984), Tamura・Masuda (2005), Tamura et al. (2006)
18	Niigata Lowland	b	○		Niigata Ancient Dune Research Group (1974), Moriwaki (1982), Nakagawa (1987), Ohira (1992), Tanaka et al. (1996), Yasui et al. (2002), Yoshida et al. (2006), Urabe et al. (2006), Urabe (2008), Urabe et al. (2011), Tanabe (2013)
19	Lake Kamo	a	△		Kobayashi et al. (1993), Hirai (1995), Nguyen et al. (1998), Matsunaga・Ota (2001)
19	Kuninaka Lowland	b	○	○	Matsunaga・Ota (2001), Ota et al. (2008), Matsubara (2009a)
20	Iwaki Lowland	c			Fujimoto (1988)
21	Lake Kasumigaura	a	○	○	Suzuki・Saito (1987), Saito et al. (1990), Hirai (1994), Sakaguchi et al. (2009), Matsubara (2009b)
22	Kujukurihama Lowland	c	○	○	Moriwaki (1979, 1982), Masuda et al. (2001 a, b), Tamura et al. (2003, 2006, 2008), Matsubara (2014)
23	Obitsu River Lowland	e		○	Kaizuka et al. (1979), Saito (1995), Matsubara (2013)
24	Futtsu Lowland	f–2	△	○	Kayane (1991)
25	Tateyama Lowland	c			Matsushima・Yoshimura (1979), Fujiwara et al. (2006), Matsubara (2013)
26	Sagami River Lowland	e	○	○	Ota・Seto (1968), Kaizuka・Moriyama (1969), Matsubara (2000a)

STUDY AREA AND METHODS OF DETERMINING RIDGE DEVELOPMENT

			A	B	
27	Matsuzaki Lowland	d-1	△		Matsubara et al. (1986), Matsubara (1988, 2000a)
28	Ukishimagahara Lowland	b	○		Matsubara (1984, 1988, 1992, 2000a), Fujiwara et al. (2007, 2008), Komatsubara et al. (2007)
29	Shimizu Lowland	c	○	○	Matsubara (1988, 1999, 2000a)
29	Miho Spit	f-1	○	○	Matsubara (1988, 1999, 2000a)
30	Haibara Lowland	d-2	○	○	Matsubara (1988, 2000a)
31	Lake Hamana	a	○		Saito (1988), Ikeya et al. (1990), Matsubara (2000a, 2001, 2004)
31	Hamamatsu Lowland	c	○	○	Matsubara (2001, 2008), Sato et al. (2011), Fujiwara et al. (2013)
32	Utsumi Lowland	d-2	△	△	Matsushima・Kitazato (1980), Maeda et al. (1983)
33	Toyama Lowland	b			Fuji (1975)
34	Lake Kahokugata	a	○	○	Fuji (1975), Kaseno et al. (1990)
35	Fukui Lowland	b		○	Fuji (1975)
36	Osaka and Kawachi Lowlands	b	○	○	Kajiyama・Ichihara (1972), Masuda et al. (2013)
37	Mihara Lowland	e			Takahashi (1982)
38	Amanohashidate	a	△		Hirai (1995), Uemura (2010)
39	Tottori Lowland	a		○	Akagi et al. (1993), Sato et al. (2013)
40	Lake Nakaumi	a		○	Mizuno et al. (1972), Tokuoka et al. (1990), Sadakata (1991), Hirai (1994)
41	Lake Shinji	a	○	○	Tokuoka et al. (1990), Hirai (1994)
41	Izumo Lowland	e	○	○	Hayashi (1991)
42	Suonada Lowlands	b, d-1,e		○	Shiragami (1983)
43	Kochi Lowland	b			Sadakata et al. (1988)
44	Tamana Lowland	e			Nagaoka et al. (1997)
45	Miyazaki Lowland	c	○		Nagaoka et al. (1991)
46	Kimotsuki Lowland	d-2	○		Nagasako et al. (1999)

a. barrier-lagoon complexes　b. sand and gravel ridge-backmarsh complexes　c. beach ridge plains
d. valley plains (d-1: barrier-backmarsh type, d-2: beach ridge type)
e. delta-beach ridge complexes　f. others (f-1: barrier spit, f-2: cuspate foreland)

A : Presence of coastal ridges during the early to mid-Holocene
B : Presence of basal landforms of coastal ridges
○ indicates the presence of coastal ridges or the presence of basal landforms of ridges
△ indicates such presence without any definite evidences

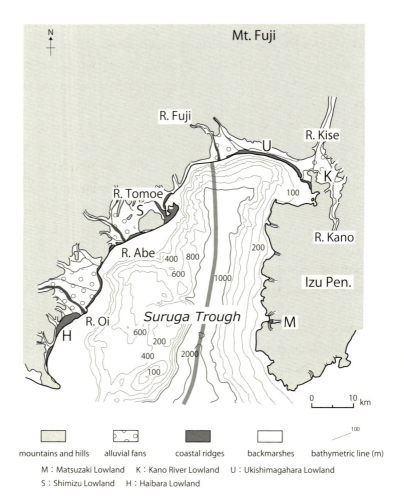

Fig. 1.1.2. Geomorphological map of coastal lowlands along Suruga Bay

delta–beach ridge complexes, and the Miho spit in the Shimizu lowland (No. 29) belongs to the barrier spit type.

Four of these areas: the Kano River and Ukishimagahara lowlands, and the Shimizu, Matsuzaki, and Haibara lowlands, are situated along Suruga Bay, which is considered to be one of the most active tectonic regions in Japan. The bay is situated at the boundary between two plates: the Philippine Sea Plate to the east and the Eurasian Plate to the west. In this region, the Philippine Sea Plate is underthrusting north-westward beneath the Eurasian Plate at the Suruga Trough (Fig. 1.1.1), which runs from north to south along the middle of the bay, with a maximum depth of more than 2,000 m (Fig. 1.1.2). It is therefore possible to clearly distinguish the influence of tectonic movements on the development of coastal ridges by studying the Suruga Bay area.

1.2. Reconstruction of Palaeoenvironmental Changes in Coastal Ridges using Fossil Foraminiferal Assemblages

Although a number of studies have been conducted on the present-day landforms of coastal ridges (Tab. 1.1.1), the methods for determining the periods during which coastal ridges were formed have been mainly based on ^{14}C dates of backmarsh deposits or the marine deposits making up the ridges. In some cases, the timing of coastal ridge formation has been estimated by the ages of archaeological sites situated upon them. These methods, however, cannot fully document the processes of coastal ridge formation through the Holocene.

To address this issue, analyses of fossil foraminiferal assemblages were conducted in this study, to provide palaeoenvironmental reconstructions of bays during the Holocene, thereby elucidating the development of coastal ridges. These analyses then allow clarification of aspects of the palaeoenvironment, such as salinity or water temperature, which could not be recog-

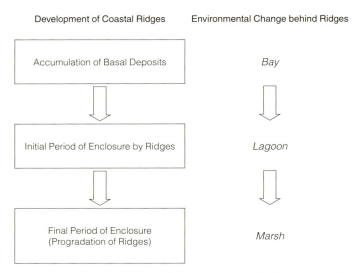

Fig. 1.2.1. Relationship between development of coastal ridges and environmental changes behind the ridges

nized from the lithology.

The relationship between the development of coastal ridges and the changes in the environment behind the ridges may be summarized as follows (Fig. 1.2.1).

(1) **Initial Stage**: Bays were formed by the Holocene transgression. During this period, the basal deposits of coastal ridges began to accumulate; however, the innermost coastal ridges had not yet emerged as coastal barriers.

(2) **Second Stage**: Coastal ridges emerged and began to enclose the bays; these bays became lagoons.

(3) **Final Stage**: The coastal ridges extended and completely enclosed the lagoons, turning them into marshes. Following this, outer coastal ridges began to develop as beach ridges seaward from the inner ridges.

Foraminifera belong to the Protozoan kingdom and consist of protoplasm enclosed by calcareous or agglutinated tests. They are extensively distributed through marine and brackish waters, and are widely spread in shal-

Tab. 1.2.1. Foraminiferal Indicators for reconstruction of palaeoenvironment

Indicators of Palaeoenvironment	Foraminiferal assemblages
Indicator A (Low salinity)	*Ammonia beccarii* forma A
Indicator B (Inflow of sea water from outside bays)	*Ammonia ketienziensis, Astrononion umbilicatulum, Bolivina robusta, Bolivina* cf. *tokiokai, Bulimina kochiensis, B.marginata, Hanzawaia nipponica, Milionlinella circularis, Pseudononion japonicum, Pseudorotalia gaimardii, Quinqueloculina seminulum, Q.vulgaris, Reussella pacifica, Triloculina trigonula*
Indicator C (Inflow of open sea water)	Planktonic Foraminifera
Indicator D (High-temperature sea water)	*Bulimina* cf. *fijiensis, Trimosina orientalis*

low and deep seas. The mean diameters of specimen tests examined in this study range from 0.1 to 1.0 mm. Fossil foraminifera are often sufficiently abundant in bore-hole cores to enable an interpretation of the environment of their deposition. Therefore, the relationship between their distribution and various physical environmental factors is quantitatively analyzed in this study.

Foraminifera are classified as planktonic or benthic. Planktonic foraminifera are used as an indicator of the distribution of sea water masses. As the distribution of modern benthic foraminifera is controlled by water depth, water temperature, salinity and sea bottom sediments, benthic foraminifera can be used as palaeo-indicators of these environmental factors.

In a bay or a drowned valley, salinity is the dominant factor controlling the distribution of benthic foraminifera, and salinity varies with the inflow of sea water from outside the bay. Therefore, changes in the inflow of sea water into an ancient bay can be recognized in analysis of fossil foraminiferal assemblages.

To set up the indicators for palaeoenvironmental reconstruction, the distribution of recent foraminifera in three different bays and lagoons is reviewed in Matsushima Bay, northeastern Japan (Matoba, 1970), Lake Hamana, central Japan (Ikeya and Handa, 1972; Ikeya, 1977), and Tanabe Bay, in southwestern Japan (Chiji and Lopez, 1968). Following this, four fo-

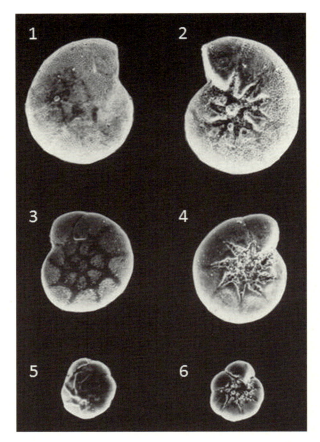

Fig. 1.2.2. Fossil foraminirera in the Holocene deposits of this study

1, 2. *Ammonia beccarii* forma A (Indicator A)   3, 4. *Ammonia beccarii* forma B   5, 6. *Ammonia beccarii* forma C   7, 8. *Pseudononion japonicum* (Indicator B)   9. *Quinqueloculina seminulum* (Indicator B)   10. Planktonic foraminifera, *Globigerina* sp. (Indicator C)   11. *Bulimina* cf. *fijiensis* (Indicator D)

raminiferal groups are established as indicators of specific palaeoenvironments: (1) Indicator A: *Ammonia beccarii* forma A, suggesting a low salinity environment such as that observed in the innermost part of a bay or near a river mouth; (2) Indicator B: the foraminiferal group that is mainly distributed near a bay mouth and indicates the inflow of sea water from outside the bay; (3) Indicator C: planktonic foraminifera, which suggest the

inflow of open sea water; and (4) Indicator D: *Bulimina* cf. *fijiensis* and *Trimosina orientalis*, which are distributed in tropical or subtropical seas (Cushman, 1942), suggesting an environment with higher water temperatures (Tab. 1.2.1, Fig. 1.2.2).

Among these, Indicator A (*A. beccarii* forma A), in particular, is useful for reconstructing the process of a bay or a drowned valley enclosure by a barrier. The initial stage of enclosure represents a change in the environment, from a bay to a lagoon. This period is therefore recognized by an increase in the frequency of *A. beccarii* forma A in the fossil foraminiferal assemblages. However, as recent foraminifera are distributed only in marine or brackish waters, the recovery of no foraminifera from the sediments suggests that the influence of the sea water has substantially decreased, which may be interpreted as an environmental change from lagoon to marsh during the final stage of enclosure by a barrier.

The analyses of fossil foraminiferal assemblages were conducted in all study areas except Amanohashidate and the Sendai and Kujukurihama lowlands.

Concerning ^{14}C data, calibrated ages are calculated using the INTCAL09/ MARINE09 (CALIB 6.01) calibration curve (Reimer *et al.*, 2009). The figures of ^{14}C ages in each area represent median probability of 1 σ calibration ranges. Regional marine reservoir correction ΔR is 200±63 yr based on Marine Data Base No. 1029, Japan, except for the Kuninaka Lowland (ΔR=50 ±30 yr; Marine Data Base No. 420, Wrangel Inlet).

Chapter 2

Geomorphic Development of Coastal Ridges in Japan

2.1. Barrier–Lagoon Complexes

2.1.1. Lake Hamana and Hamamatsu Lowland

Lake Hamana with an area of approximately 65 km^2 facing the Pacific Ocean, is a typical barrier–lagoon complex in Japan (Figs. 1.1.1; 2.1.1, 2.1.2). It is surrounded by the Palaeozoic–Mesozoic mountains to the north, and Pleistocene uplands to the east and west. Three coastal ridges are recognized at the mouth of the lake, and are numbered I to III in a seaward direction. According to historical documentation, the present barrier —coastal ridge III— blocked up the lagoon and has been broken repeatedly by tsunamis. The present mouth of the lake was artificially cut in 1956.

According to bathymetric mapping, the depth of the lake increases towards the inner portions, with the depth in this region ranging from 9 to 12

CHAPTER 2

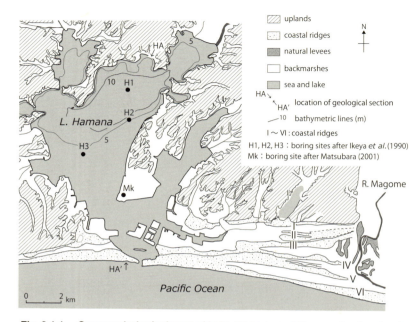

Fig. 2.1.1. Geomorphological map of Lake Hamana and Hamamatsu Lowland

Fig. 2.1.2. Lake Hamana (Front. 1)

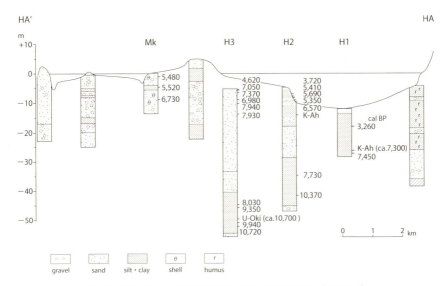

Fig. 2.1.3. Geological section of Lake Hamana (HA-HA')

m. Meanwhile, the outer parts of the lake are less than 4 m deep (Fig. 2.1.1). Lake deposits comprise muddy sediments in the inner part, and sandy sediments belonging to a flood tidal delta in the outer part (Fig. 2.1.3).

The Hamamatsu lowland, located to the east of Lake Hamana, is distributed along the southern foot of the Pleistocene uplands. Six coastal ridges are developed within this lowland and numbered I to VI in a seaward direction. The inner coastal ridges are developed on buried abrasion platforms approximately 10 m below mean sea level, in front of the former sea cliffs. The recent deposits in the lowland are mainly composed of sandy sediments forming the coastal ridges (Figs. 2.1.4, 2.1.5), whereas the surface deposits in the marshes between the coastal ridges are peaty silt and clay.

Sampling of the Holocene deposits was conducted at four locations on the lake floor by Ikeya et al. (1990). Fossil foraminifera of continuous borehole cores from locations H2 and H3 were analyzed; the depth of the cores was 5 m (Fig. 2.1.1).

Fig. 2.1.4. Excavation using a Geoslicer in Hamamatsu Lowland (Front. 3)

Fig. 2.1.5. Deposits of coastal ridge in Hamamatsu Lowland (Front. 4)

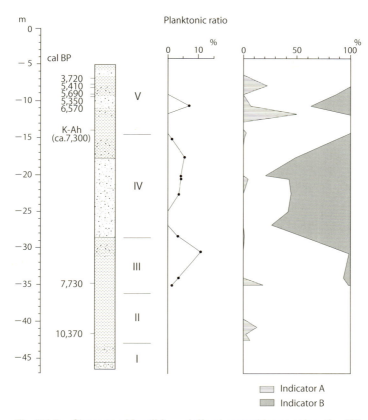

Fig. 2.1.6. Changes of fossil foraminiferal assemblages at location H2

At location H2, seven ^{14}C dates were obtained, and K-Ah volcanic glass dating to ca. 7,300 cal BP was found at −14.00 m. Twenty-eight samples were analyzed from the bore-hole cores between −46.4 and −5.0 m, with an estimated maximum age of ca. 10,900 cal BP. Fossil foraminifera were observed in the deposits from −42.9 to −5.0 m.

The sediment is divided into five parts, numbered upward from I to V, according to their fossil foraminiferal assemblages as follows (Fig. 2.1.6).

I (−46.4 to −42.9 m, 10,900 to 10,500 cal BP)

No foraminifera were found; therefore, it is considered that the sea water had not yet invaded the region.

II (−42.9 to −36.2 m, 10,500 to 9,800 cal BP)

Elphidium advenum was dominant, accompanied by *Ammonia. beccarii* forma A, B, and C, and *Elphidium. reticulosum*. Neither Indicator B, suggesting the inflow of sea water from outside the bay, nor planktonic foraminifera, suggesting the inflow of open sea water, were observed. Consequently, it is inferred that the sea water began to invade the region around 10,500 cal BP.

III (−36.2 to −28.6 m, 9,800 to 8,900 cal BP)

The frequencies of *A. beccarii* forma B and C and *E. advenum* increased. Furthermore, both Indicator B and planktonic foraminiferal assemblages were observed. These suggest that the sea water from outside the bay and open sea water began to invade the bay at this time.

IV (−28.6 to −14.6 m, 8,900 to 7,300 cal BP)

Pseudononion japonicum, one of the Indicator B species, became dominant, whereas the proportion of planktonic foraminifera was approximately 5%. As the sediments consist of sand, from which the coastal ridges are constructed, the sea water is considered to have invaded and deposited sandy materials in the bay.

V (−14.6 to −5.0 m, since ca. 7,300 cal BP)

The frequency of *A. beccarii* forma A increased, indicating a low salinity environment, whereas that of Indicator B species decreased. Furthermore, *Trochammina* spp. and *Haplophragmoides* spp. were observed. These are characterized as having agglutinated tests and living in a low salinity environment. Consequently, it is considered that the bay began to be enclosed by the coastal barrier, and was transformed into a lagoon at this time.

At location H3, ten ^{14}C dates were obtained, and U-Oki volcanic ash dating to ca. 10,700 cal BP was found at −49.6 m. Forty-seven samples were analyzed from the bore-hole cores between −55.0 and −5.5 m. Fossil foraminifera were observed in the deposits from −52.4 to −10.0 m, spanning an estimated period from 10,500 to 7,300 cal BP.

The sediments are divided into eight parts, numbered upward from I to

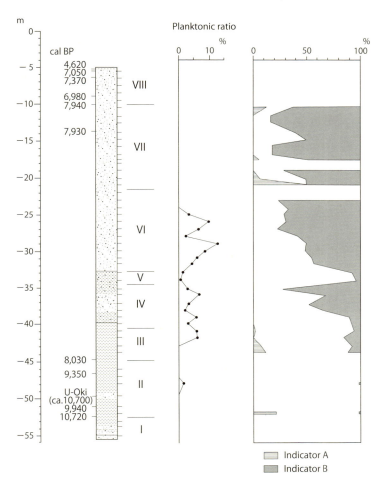

Fig. 2.1.7. Changes of fossil foraminiferal assemblages at location H3

VIII, according to fossil foraminiferal assemblages as follows (Fig. 2.1.7).

I (−55.0 to −52.4 m, 10,900 to 10,500 cal BP)

No foraminifera were found; therefore, it is inferred that the sea water had not yet invaded the region at this time.

II (−52.4 to −44.8 m, 10,500 to 9,500 cal BP)

Fossil foraminifera were found in the deposits at −51.9 and −48.0 m. *A.*

beccarii forma B and C were dominant, whereas the Indicator B species scarcely occurred, and planktonic foraminifera were only found at −48.0 m. Therefore, it is considered that the sea water began to invade the region around 10,500 cal BP.

III (−44.8 to −40.4 m, 9,500 to 9,000 cal BP)

A. beccarii forma A, B, and C, *E. advenum* and *P. japonicum* were dominant in this section, and frequencies of Indicator B and planktonic foraminiferal assemblages also increased. This suggests that both sea water from outside the bay and open sea water had invaded the bay at this time.

IV (−40.4 to −34.5 m, 9,000 to 8,100 cal BP)

A. beccarii forma B and C and *P. japonicum* (one of the Indicator B species) were dominant; however, *A. beccarii* forma A, which indicates a low salinity environment, was not observed. Consequently, the influence of sea water from outside the bay is considered to have increased. Additionally, the sandy deposits in this section were coarser than those in II and III, suggesting that the coastal barrier had begun to be formed in this period.

V (−34.5 to −32.8 m, 8,100 to 8,000 cal BP)

A. beccarii forma B and C, *Buliminella elegantissima* and *Elphidium somaense* were dominant, whereas the frequency of Indicator B and planktonic foraminiferal assemblages decreased. It is therefore inferred that the influence of the sea water decreased for a while during this period.

VI (−32.8 to −21.5 m, 8,000 to 7,700 cal BP)

P. japonicum was dominant, and the frequency of Indicator B and planktonic foraminiferal assemblages increased again within the deposits. Consequently, the sea water is considered to have invaded and supplied sandy deposits into the bay, allowing construction of the coastal barrier.

VII (−21.5 to −10.0 m, 7,700 to 7,300 cal BP)

Although *P. japonicum* was dominant, *A. beccarii* forma A, indicating a low salinity environment, also occurred. This suggests that the coastal barrier had begun to enclose the bay at this time.

VIII (−10.0 to −5.5 m, since ca. 7,300 cal BP)

As no foraminifera were observed in this section, it is inferred that the coastal barrier had developed and began to emerge.

At the location of Mk, around 4 km inland from the mouth of the lake and situated within the inner part of the present coastal barrier (Fig. 2.1.1), three ^{14}C dates were obtained from shell fragments in the sandy deposits: 6,730 cal BP (−8.3 m), 5,520 cal BP (−4.3 m) and 5,480 cal BP (−0.3 m). These results indicate that the coastal barrier systems had already developed by the time the Holocene transgression culminated.

The overall geomorphic development of Lake Hamana can therefore be summarized as follows, according to the results from both H2 and H3 (Fig. 2.1.8).

The bay started to form by around 10,500 cal BP, with sea water exerting the most influence on the bay between 9,000 and 8,000 cal BP. During this period, the barrier system began to develop, and when coastal ridge I —formerly the coastal barrier— began to enclose the bay around 7,500 cal BP, the bay was changed into a lagoon.

The sedimentation rate at both H2 and H3 decreased after 5,000 to 4,000 cal BP. This indicates that coastal ridge I had completely enclosed the bay by this time, and the outer coastal ridges began to develop seaward after ca. 5,000 cal BP.

In the Hamamatsu lowland, construction of coastal ridge I was complete before 7,000 cal BP as indicated by the ^{14}C date of ca. 6,960 cal BP of the peaty clay just above the coastal ridge I deposits (Institute of Palaeoenvironmental Resaerch, 1994). The final stage of enclosure for the area behind coastal ridge I in the western part of the Hamamatsu lowland was inferred to have occurred between 3,400 and 3,200 cal BP, indicated by ^{14}C dates obtained from the bottom of the peaty sediment (Sato, *et al.*, 2011).

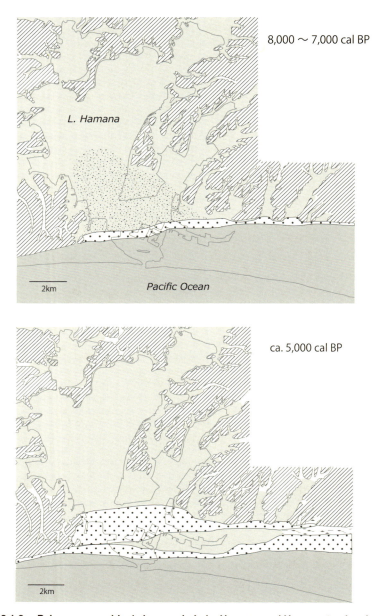

Fig. 2.1.8. Palaeogeographical changes in Lake Hamana and Hamamatsu Lowland

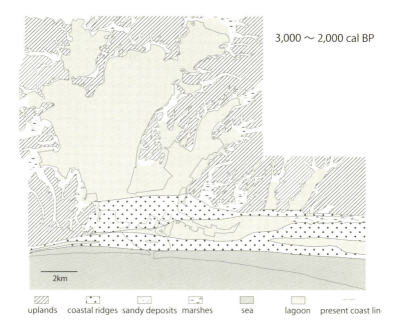

2.1.2. Amanohashidate

Amanohashidate is one of the most famous coastal barriers in Japan; it is situated in the southeastern part of the Tango Peninsula in Kyoto Prefecture, facing the Sea of Japan (Figs. 1.1.1; 2.1.9, 2.1.10). Amanohashidate developed from north to south, approximately 3 km in length. It encloses Miyazu Bay, which is one of the inlets along Wakasa Bay. The lagoon behind the coastal barrier is called Aso-Kai (Lake Aso). According to the bathymetric map of Lake Aso, the deepest area around −13 m is situated just behind the barrier (Fig. 2.1.9). The lake deposits are mainly composed of muddy sediment (Hirai, 1995). The Noda River flows into Lake Aso at the western (innermost) part of the lake. Coastal ridges develop on the delta around the mouth of the Noda River.

According to Uemura (2010), the Holocene marine deposits along the Noda River are recognized in the area about 1.5 km inland from the river mouth. This suggests that a former bay expanded around Lake Aso during

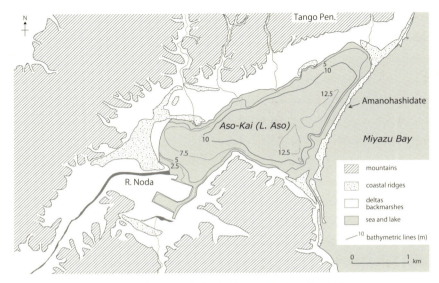

Fig. 2.1.9. Geomorphological map of Amanohashidate

Fig. 2.1.10. Amanohashidate from the northern mountains (Front. 6)

Miyazu Bay (left side), Lake Aso (right side)

the Holocene transgression. Matsubara and Shiomi (2010) analyzed the fossil molluscan assemblages in the Holocene marine deposits obtained in the Noda River delta. The fossil molluscan assemblages mainly consist of *Ringicula doliaris* and *Papiha undulate* belonging to a typical assemblage found in the middle part of the bay with muddy bottoms. Consequently, it is inferred that the palaeoenvironment of the inner part of Lake Aso had not been influenced by the open sea water.

2.2. Sand and Gravel Ridge–Backmarsh Complexes

2.2.1. Kano River and Ukishimagahara Lowlands

The Kano River and Ukishimagahara lowlands are developed along the head of Suruga Bay, which is surrounded by Quaternary volcanoes such as Mt. Fuji, Mt. Ashitaka, and Mt. Hakone to the north, and volcanic groups of the Izu Peninsula to the southeast (Figs. 1.1.1, 1.1.2; 2.2.1, 2.2.2).

The Kano River rises on Mt. Amagi, in the central part of the Izu Peninsula and flows into Suruga Bay, while the Kise River, a tributary of the Kano River, flows from the eastern foot of Mt. Fuji. The Kano River lowland includes a valley plain in its middle to lower reaches, a volcanic fan in the lower reaches of the Kise River and an alluvial delta at the mouth of the Kano River. The recent deposits in the valley plains comprise basal sand and gravel, followed by lower fluvial deposits, a middle clay unit, upper pumiceous sands and gravels, topped by alluvial deposits (Fig. 2.2.3). The basal sand and gravel is considered to represent the deposits of the former river bed. The middle clay unit was deposited under a low salinity environment, and is divided into two parts: lower marine clay and upper brackish clay. The pumice (Kg) contained in the pumiceous sand and gravel was derived from the pumice fall and flow deposits ejected around 3,100 cal BP from the Kawagodaira volcano in the upper reaches of the Kano River.

Holocene deposits were obtained for analysis at the K1 location, 10.04 m

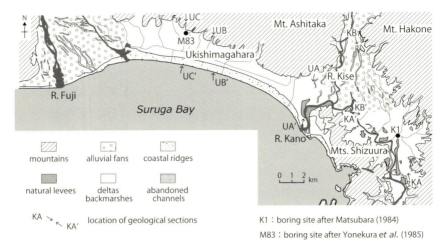

Fig. 2.2.1.　Geomorphological map of Kano River and Ukishimagahara Lowlands

Fig. 2.2.2.　Ukishimagahara Lowland at the southern foot of Mt. Fuji (Front. 7)

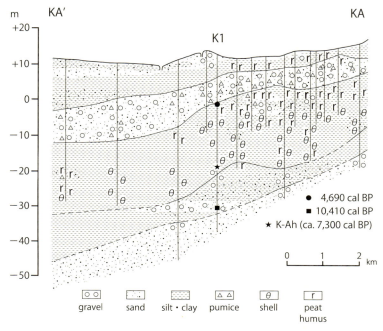

Fig. 2.2.3. Geological section of Kano River Lowland (KA-KA')

above mean sea level, in the middle part of the lowland (Fig. 2.2.1). Samples were taken from discontinuous bore-hole cores at intervals of around 1 m. Two ^{14}C dates (4,690 cal BP at −1.33 m and 10,410 cal BP at −30.88 m) were obtained, and K-Ah volcanic glass dating to ca. 7,300 cal BP was found in the deposits at −18.9 m. Twelve samples from the bore-hole cores were analyzed for foraminifera, between −24.9 and −10.9 m, spanning an estimated period from 8,800 to 6,100 cal BP. Fossil foraminifera were observed in all samples. In general, the frequency of the Indicator B assemblage, suggesting the inflow of sea water from outside the bay, was larger than that of the Indicator A species, *Ammonia beccarii* forma A, which indicates a low salinity environment. Planktonic foraminifera, suggesting the inflow of open sea water, occurred in the deposits between −23.9 and −14.9 m (8,500 to 6,700 cal BP), whereas the Indicator A assemblage became dominant in the deposits between −14.9 and −10.9 m (ca. 6,700 to 6,100 cal BP). Following this, backmarsh deposits were observed to accumulate above −10.9 m (since ca. 6,100 cal BP).

These results suggest that the bay that was formed in the Kano River lowland was changed into a lagoon around 6,500 cal BP, and this lagoon transformed into a backmarsh around 6,000 cal BP. This enclosure was caused by the development of a coastal barrier at the mouth of the bay.

The volcanic fan known as the Mishima fan consists of Mishima lava that flowed from Mt. Fuji around 12,300 cal BP, in addition to fan gravel covering the lava. Marine sands are distributed around the southern end of the fan. The Mishima lava unit abuts the foot of the Shizuura mountains to the south, and a narrow pass is formed between the lava and the rocks of the Shizuura mountains (Fig. 2.2.4). The alluvial delta at the mouth of the Kano River is composed, from the bottom to the top, of marine gravel, humic clay, and deltaic sand.

The Ukishimagahara lowland, representing a sand and gravel–backmarsh complex (Fig. 2.2.1), is located to the west of the Kano River delta, and adjoins the alluvial fans of the Fuji River to the west. The lowland extends approximately 20 km in length from east to west, and is 2 km wide from north to south. The coastal ridges in this region comprise sand and gravel mainly supplied from the Fuji River. In the backmarsh area, peaty clay with scoria (ObS) is underlain by marine sand and gravel. The surface height of the marine sand and gravel deposits, including the buried barrier (coastal ridge I) deposits, becomes deeper towards the northwest (both toward the Suruga Trough and landward). Coastal ridge II, which is buried behind the present-day coastal ridge III, had prograded seaward from coastal ridge I in the eastern and central parts of the Ukishimagahara lowland, whereas in the western portion of the lowland, the ridge developed upward. This is likely caused by local differences in the subsidence rate (Figs. 2.2.5, 2.2.6, 2.2.7, 2.2.8, 2.2.9).

Sampling of the Holocene deposits was conducted at location M83, 4.2 m

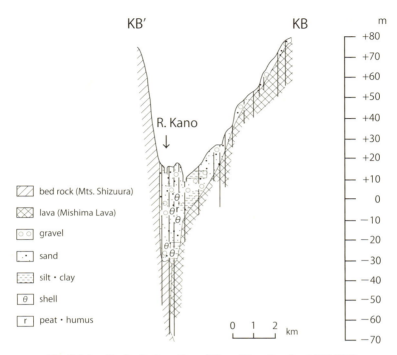

Fig. 2.2.4. Geological section of Kano River Lowland (KB-KB')

above mean sea level and situated in the innermost part of the backmarsh behind the present-day coastal ridge by Yonekura *et al.* (1985). Seven ^{14}C dates were obtained, and K-Ah volcanic glass (ca. 7,300 cal BP) was recovered from −18.8 m. Forty-eight samples from the continuous bore-hole cores, between −40.8 and −12.8 m were analyzed for foraminifera. Fossil foraminifera were observed in the deposits from −40.5 to −15.2 m, spanning an estimated period from 10,500 to 5,800 cal BP. The sediments may be divided into seven parts, numbered upward from I to VII, according to their fossil foraminiferal assemblages as follows (Fig. 2.2.10).

I (−40.5 to −39.3 m, ca. 10,000 cal BP)

A. beccarii forma A was dominant, indicating a low salinity environment. This suggests that the sea water began to invade the Ukishimagahara lowland around 10,000 cal BP.

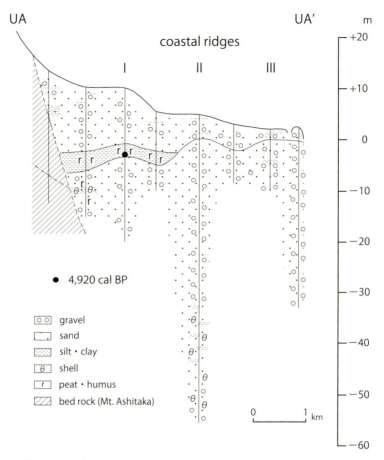

Fig. 2.2.5. Geological section of Ukishimagahara Lowland (UA-UA')

II (−39.3 to −32.8 m, 10,000 to 9,000 cal BP)

No foraminifera were observed except for in one sample at −36.5 m. As the sediments contain fluvial sand with scoria, it is inferred that the sea water had only a slight influence for a short period.

III (−32.8 to −24.8 m, 9,000 to 8,000 cal BP)

A. beccarii forma B and C, *Buliminella elegantissima*, *Elphidium subgranulosum*, *E. advenum*, *E. jenseni*, and *E. reticulosum* were found, which belong to

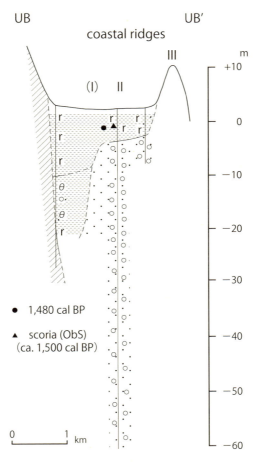

Fig. 2.2.6. Geological section of Ukishimagahara Lowland (UB-UB')

the typical foraminiferal assemblage found in a bay. The Indicator B species, suggesting the inflow of sea water from outside the bay, also occurred. Furthermore, planktonic foraminifera indicating the inflow of open sea water, and *Bulimina* cf. *fijiensis*, suggesting an environment with a higher water temperature, were found at −28.0 m, indicating the influence of sea water from outside the bay around 8,200 cal BP.

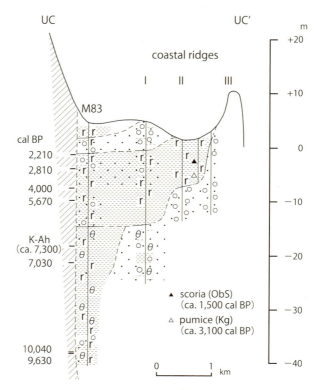

Fig. 2.2.7. Geological section of Ukishimagahara Lowland (UC-UC')

IV (−24.8 to −22.3 m, 8,000 to 7,700 cal BP)

No foraminifera were found except for in the sample at −24.5 m. As the sediments contain coarse sand with plant remains, it is considered that the inland water flowed into the bay during this period.

V (−22.3 to −19.8 m, 7,700 to 7,400 cal BP)

A. beccarii forma A became dominant, indicating the beginning of enclosure of the bay by the barrier, transforming it into a lagoon.

VI (−19.8 to −15.2 m, 7,400 to 6,800 cal BP)

The fossil foraminiferal assemblage mainly consisted of *A. beccarii* forma A, indicating that the effects of the barrier enclosure have increased.

Fig. 2.2.8. Buried coastal ridge in Ukishimagahara Lowland (Front. 9)
Sandy deposits of former coastal barrier were observed under the peat. The depth of trench from the land surface is around 2 m.

Fig. 2.2.9. Pumice and Scoria in backmarsh deposits of Ukishimagahara Lowland
Kawagodaira Pumice (Kg) (ca. 3,100 cal BP) (left), Obuchi Scoria (ObS) (ca. 1,500 cal BP) (right)

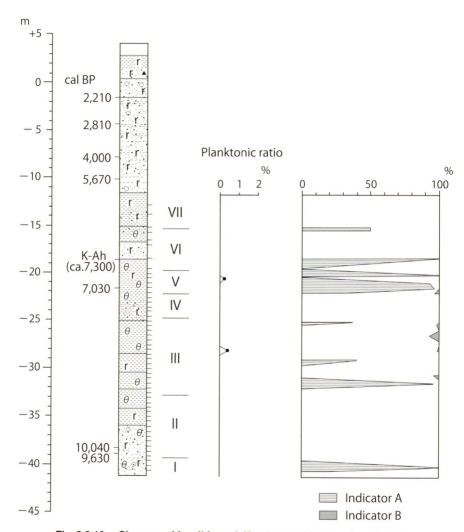

Fig. 2.2.10. Changes of fossil foraminiferal assemblages at location M83

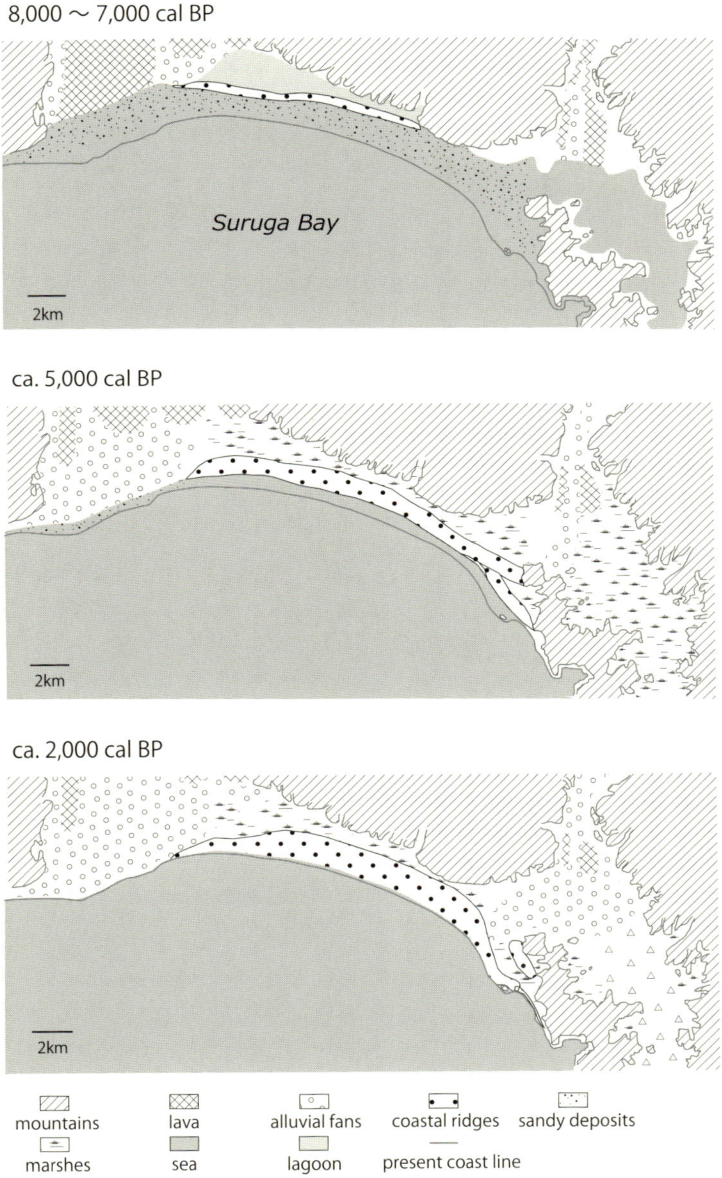

Fig. 2.2.11. Palaeogeographical changes in Kano River and Ukishimagahara Lowlands

VII (above −15.2 m, since ca. 6,800 cal BP)

No foraminifera were found, and the sediments contained peat; therefore, the lagoon is inferred to have turned into a marshy area after ca. 6,800 cal BP.

According to these results, the geomorphic development of the Ukishimagahara lowland during the Holocene may be summarized as follows (Fig. 2.2.11).

Sea water that invaded the region began to form a bay around 10,000 cal BP. Although the influence of sea water from outside the bay increased between 9,000 and 8,000 cal BP, open sea water had not yet directly influenced the environment of the bay at this time. Around 8,000 to 7,400 cal BP, the bay began to be enclosed by a coastal barrier and was transformed into a lagoon. This lagoon then turned into a marsh since after ca. 6,800 cal BP. As marine sand and gravel had been already deposited between 9,000 and 8,000 cal BP, coastal ridge I must have begun to form before ca. 8,000 cal BP, emerging between 8,000 and 7,400 cal BP and completely enclosing the lagoon by ca. 6,800 cal BP. This is supported by the results of Fujiwara *et al.* (2008), based on the analyses of sedimentary facies, fossil ostracodes and molluscan assemblages of the Holocene deposits behind coastal ridge I. The peat behind coastal ridge II, which developed seaward from coastal ridge I, began to accumulate around 5,000 cal BP (Fig. 2.2.5); consequently, coastal ridge II is considered to have enclosed the backmarsh around this time. Furthermore, the present-day coastal ridge III is considered to have been constructed before 2,000 cal BP, because historic relics found both on the ridge and in the backmarsh date from the Yayoi period, around 2,000 cal BP.

2.2.2. Tokoro Lowland

The Tokoro lowland faces the Sea of Okhotsk, and is surrounded by the Neogene hills to the east and Quaternary uplands to the west. The Tokoro River is located in the eastern part of the lowland, and flows into the Sea of

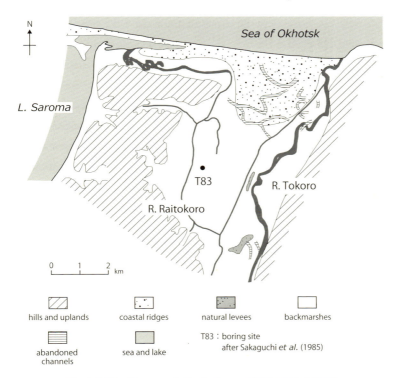

Fig. 2.2.12. **Geomorphological map of Tokoro Lowland**

Okhotsk, while the Raitokoro River in the western part of the lowland flows into Lake Saroma (Figs. 1.1.1; 2.2.12). Barrier–lagoon complexes are developed in Lake Saroma to the west of the Tokoro lowland, with a coastal barrier extending toward the Tokoro lowland. A former flood tidal delta has been recognized behind the present-day coastal ridge (Saito, 1987; Matsubara, 2000a).

Sampling of Holocene deposits was conducted in the backmarsh, at location T83, which is 3.0 m above mean sea level, by Sakaguchi et al. (1985). The recent deposits in the lowland comprise, from the bottom to the top, sand and gravel, silt with plant remains, clay with shell fragments and peaty clay. Twelve ^{14}C dates were obtained.

Forty-three samples were obtained from the continuous bore-hole cores between −31.5 and −4.2 m, and were analyzed for foraminifera. Fossil foraminifera were observed in the deposits from −28.1 to −12.3 m, spanning an estimated period from 9,900 to 6,100 cal BP.

The sediments may be divided into six parts, numbered upward from I to VI, on the basis of their fossil foraminiferal assemblages as follows (Fig. 2.2.13).

I (−31.5 to −28.1 m, 10,400 to 9,900 cal BP)

No foraminifera were found; therefore, it is considered that the sea water had not yet invaded the area at this time.

II (−28.1 to −22.6 m, 9,900 to 8,400 cal BP)

Ammonia beccarii forma A, B, and C, *Elphidium*, *subincertum*, and *Buccella frigida* were dominant. *B. frigida* indicates an environment with low water temperature. Planktonic foraminifera, suggesting the inflow of open sea water, were not found. Consequently, it is supposed that a bay began to expand during this period; however, the influence of sea water from outside the bay was, as yet, quite small.

III (−22.6 to −20.7 m, 8,400 to 7,900 cal BP)

Although the benthic foraminiferal assemblages were the same as in section II, planktonic foraminifera began to be observed in these deposits. This suggests that open sea water began to invade the bay around 8,000 cal BP.

IV (−20.7 to −18.3 m, 7,900 to 7,400 cal BP)

No foraminifera were observed; therefore, it is considered that only inland water flowed into the bay during this period.

V (−18.3 to −12.3 m, 7,400 to 6,100 cal BP)

The total number of foraminiferal species decreased. *A. beccarii* forma A, B and C were dominant, and frequency of *A. beccarii* forma A increased upwards. No planktonic foraminifera were observed in this section. Consequently, it is considered that the bay began to transform into a lagoon during this period.

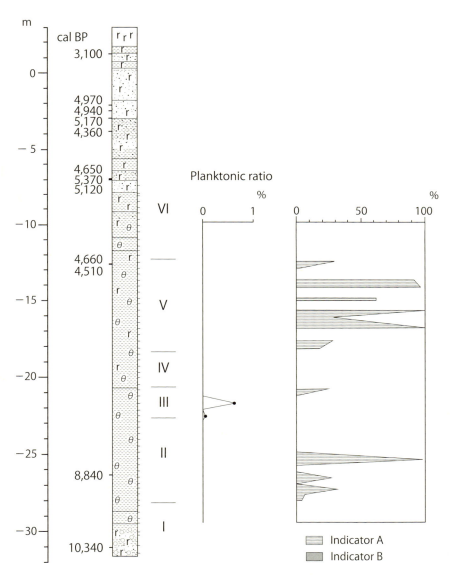

Fig. 2.2.13. Changes of fossil foraminiferal assemblages at the location T83

VI (above −12.3 m, since ca. 6,100 cal BP)

No foraminifera were found, implying that the coastal ridge had begun to emerge, and the lagoon was transformed into a marsh around 6,100 cal BP.

Peaty clay was observed above −9.0 m (since ca. 5,400 cal BP). This suggests that the coastal ridge had completely enclosed the backmarsh by around 5,400 cal BP.

From these results, the geomorphic development around location T83 during the Holocene can be summarized as follows.

The bay was expanding between 9,900 and 7,900 cal BP. Subsequently, the bay changed into a lagoon around 7,900 cal BP, as a coastal ridge began to enclose the bay. The coastal ridge had completely enclosed the lagoon by around 5,400 cal BP.

2.2.3. Kuninaka Lowland

The Kuninaka lowland is situated in the central part of Sado Island in the Sea of Japan, which is within one of the most seismologically active areas in Japan. The lowland is located between two mountainous areas, the Osado mountains to the northwest and the Kosado mountains to the southeast (Figs. 1.1.1; 2.2.14). The northeast-trending Kuninaka-minami fault represents an approximate boundary between the Kuninaka lowland and the Kosado mountains (Ota *et al.*, 2008). Furthermore, the Kuninaka lowland faces the Mano Bay, with the Kokufu River flowing into Mano Bay. Coastal ridges are developed along both Ryotsu and Mano Bays. The coastal ridges along Mano Bay belong to barrier–backmarsh complexes in valley plains, whereas the ridges developed along Ryotsu Bay can be characterized as barrier–lagoon complexes. Lake Kamo is situated behind the coastal ridges.

The average height of the coastal ridges along Mano Bay is around 5 m above mean sea level. However, the maximum height of the ridges is around 20 m on both sides of the Kokufu River, where the Pleistocene terraces at the northern end of the Kosado mountains are buried in the lowland

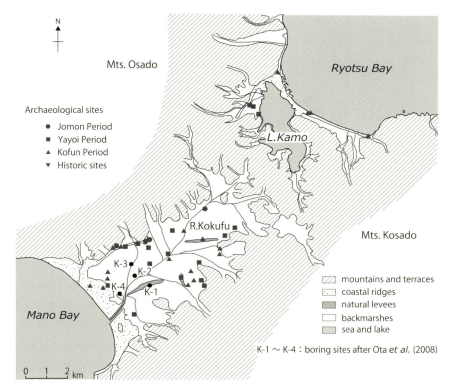

Fig. 2.2.14. Geomorphological map of Kuninaka Lowland

and recognized as abrasion platforms. This indicates that the development of the coastal ridges is affected by their basal landforms. In the central part of the Kuninaka lowland, natural levees are developed along the Kokufu River.

Sampling of the Holocene deposits in this area was conducted at four locations (K-1 to K-4) (Ota et al., 2008) (Fig. 2.2.15). Fifteen ^{14}C dates were obtained, and foraminiferal assemblages were analyzed as follows.

K-1

Six samples were analyzed for foraminifera from the continuous borehole cores at location K-1, at a height of 4.0 m above mean sea level. Fossil foraminifera were recovered from the Holocene deposits from −11.6 to −5.5 m, spanning an estimated period from 8,400 to 6,700 cal BP. It is in-

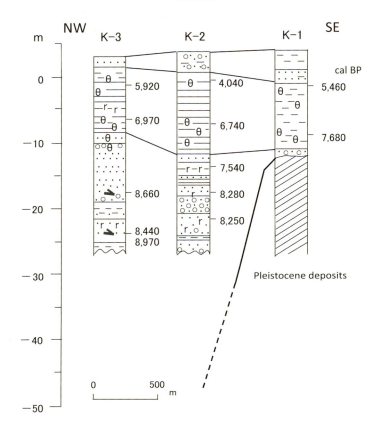

Fig. 2.2.15. Geological sections of Kuninaka Lowland

ferred that the deposits below −11.6 m are the Pleistocene marine sediments making up the buried terrace according to ^{14}C dates.

The sediments may be divided into three parts based on the fossil foraminiferal assemblages, numbered upward from I to III as follows.

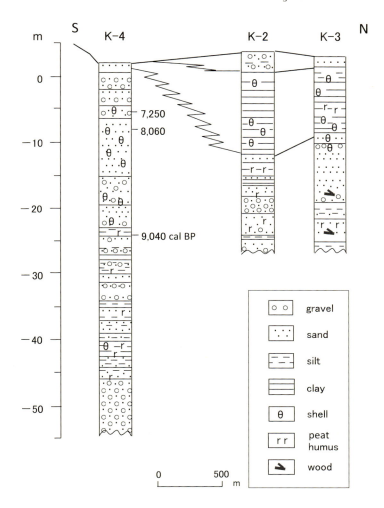

I (−11.6 to −8.5 m, 8,400 to 7,500 cal BP)

The total number of observed fossils was low, suggesting that the influence of sea water was not very large during this period.

II (−8.5 to −6.6 m, 7,500 to 7,000 cal BP)

Ammonia. beccarii forma B and C and *Buccella frigida* were dominant. A.

beccarii belongs to a typical foraminiferal assemblage found in a bay, and *B. frigida* indicates an environment with low water temperature. The propotion of planktonic foraminifera, which indicate the inflow of open sea water, was 10 % to 20 %. Consequently, the inflow of sea water from a cold current was significant during this period.

III (−6.6 to −5.5 m, 7,000 to 6,700 cal BP)

A. beccarii forma A was dominant, indicating a low salinity environment. However, neither the Indicator B assemblage, suggesting the inflow of sea water from outside the bay, nor planktonic foraminifera were found. It is therefore inferred that a coastal barrier began to enclose the bay, and that the bay changed into a lagoon at this time.

K-2

Nine samples were analyzed for foraminifera from the continuous borehole cores at location K-2, at a height of 3.7 m above mean sea level. Fossil foraminifera were observed in the deposits from −10.7 to −6.2 m, spanning an estimated period from 7,100 to 6,300 cal BP.

The sediment may be divided into three parts according to their fossil foraminiferal assemblages and numbered upward from I to III as follows.

I (−10.7 to −9.3 m, 7,100 to 7,000 cal BP)

A. beccarii forma A was dominant, indicating that the influence of sea water from outside the bay was small.

II (−9.3 to −8.5 m, 7,000 to 6,900 cal BP)

The frequency of *A. beccarii* forma A decreased, whereas *Elphidium somaense* and *E. subincertum*, belonging to the typical foraminiferal assemblage found in a bay, increased in number. This suggests that the inflow of sea water from outside the bay increased.

III (−8.5 to −6.2 m, 6,900 to 6,400 cal BP)

The Indicator B assemblage (suggesting the inflow of sea water from outside the bay) occurred in the deposits, indicating that the influence of sea water became larger.

K-3

Eleven samples were analyzed for foraminifera from the continuous bore-hole cores at location K-3, at a height of 3.0 m above mean sea level. Fossil foraminifera were observed in the deposits from −7.8 to −0.3 m, spanning an estimated period from 7,800 to 5,200 cal BP.

The sediment is divided into four parts, numbered upward from I to IV, according to their fossil foraminiferal assemblages as follows.

I (−7.8 to −5.6 m, 7,800 to 7,100 cal BP)

A. beccarii forma B and C were dominant, accompanying *B. frigida* and *A. beccarii* forma A (Inidicator A), *Pseudononion japonicum* and *Quinqueloculina seminulum*, which are Indicator B species suggesting the inflow of sea water from outside the bay, also occurred in the deposits. Together, these suggest that sea water invaded the bay during this period.

II (−5.6 to −4.3 m, 7,100 to 6,600 cal BP)

A. beccarii forma A, indicating a low salinity environment, accounts for over 90 % of the total foraminifera; therefore, it is inferred that the influence of sea water from outside the bay decreased for a period.

III (−4.3 to −2.1 m, 6,600 to 5,900 cal BP)

The frequency of *A. beccarii* forma A decreased, while *Q. seminulum*, which belongs to the Indicator B assemblage, was observed. This suggests that the inflow of sea water from outside the bay once again increased.

IV (−2.1 to −0.3 m, 5,900 to 5,200 cal BP)

The influence of sea water outside the bay decreased again, indicated by an increase in the frequency of *A. beccarii* forma A.

K-4

Nineteen samples were analyzed for foraminifera from the continuous bore-hole cores at location K-4, at a height of 2.0 m above mean sea level. Fossil foraminifera were found from −35.5 to −4.8 m, spanning an estimated period from 9,800 to 7,100 cal BP.

The sediment may be divided into three parts according to their fossil foraminiferal assemblages, numbered upward from I to III as follows.

CHAPTER 2

I (−35.5 to −22.0 m, 9,800 to 8,900 cal BP)

No foraminifera were found; therefore, it is considered that the sea water had not yet invaded the area.

II (−22.0 to −19.0 m, 8,900 to 8,700 cal BP)

Elphidium subarcticum, belonging to the typical foraminiferal assemblage found in a bay, was dominant. The proportion of planktonic foraminifera, indicating the inflow of open sea water, was over 20 %. These together suggest that the bay was expanding as a result of the influence of open sea water during this period.

III (−19.0 to −4.8 m, 8,700 to 7,100 cal BP)

Cibicides spp., *Elphidium crispum*, and *E. subarcticum*, belonging to the typical foraminiferal assemblage found in a bay, were observed. Furthermore, *P. japonicum* (one of the Indicator B species, suggesting the inflow of sea water from outside the bay) also occurred alongside planktonic foraminifera with a frequency of over 40 %. Consequently, the influence of open sea water is considered to have increased during this period.

Among the four sites in this area, location K-4 is nearest to the sea. As the deposits at the K-4 site are mainly composed of sand, the site is considered to represent the inner part of the coastal ridge.

The characteristics of the fossil foraminiferal assemblages at K-4 are different from those at the three other sites (K-1, K-2, and K-3). *A. beccarii* forma A is dominant at the other sites; however, it was not found at K-4. In contrast, *E. crispum* and *Cibicides* spp. adhered to rocks were dominant in the latter site, while planktonic foraminifera, indicating the inflow of open sea water, also occurred. This suggests that the influence of open sea water was greater at K-4 than at the inner sites. In this case, *E. crispum* and *Cibicides* spp. are considered to be allochthonous deposits supplied from the Pleistocene sediments around the bay.

As marine sand and gravel had already been deposited around 9,000 cal BP at K-4, the coastal ridge began to form during the stage of sea level rise in the early Holocene. According to the above analyses at locations K-1 and

K-3, the bay began to be enclosed by the coastal ridge and transformed into a lagoon around 7,000 to 6,000 cal BP. This period corresponds to the stable stage in the Holocene relative sea level change in the Kuninaka lowland according to Ota *et al.* (2008). Furthermore, the lagoon is considered to have become a marsh around 5,000 cal BP, according to analyses of fossil diatoms at K-1 (Ota *et al.*, 2008).

From the results obtained at K-1 to K-4, the geomorphic development of the Kuninaka lowland during the Holocene may be summarized as follows.

Coastal ridges began to form during the stage of sea level rise around 9,000 cal BP and developed on the buried abrasion platforms. The bay began to be enclosed by the coastal ridges during 7,000 to 6,000 cal BP, corresponding to a stable stage in the Holocene relative sea level change in the Kuninaka lowland.

2.2.4. Sendai Lowland

The Sendai lowland faces the Pacific Ocean, extends around 50 km in length from north to south, and is 10–20 km wide from east to west (Fig. 1.1.1). The Nanakita, Natori, Hirose, and Abukuma Rivers flow through the lowland into Sendai Bay. Most parts of the lowland have an altitude of less than 5 m, and the landforms in this area are characterized by both natural levees along the rivers and three coastal ridges, which are numbered I to III in a seaward direction (Fig. 2.2.16).

The present-day landforms and the deposits in the lowland have been analyzed by Matsumoto (1981, 1984), Tamura and Masuda (2005), and Tamura *et al.* (2006). Among these, Matsumoto (1981, 1984) established the geomorphic development of the coastal ridges during the Holocene on the basis of ^{14}C dates.

According to their results, coastal ridges I, II, and III in the northern part of the lowland were constructed around 5,000 cal BP, 2,500 cal BP, and 700 cal BP, respectively. In the central part of the lowland, coastal ridges I, II, and III developed during 4,500 to 4,000 cal BP, 3,300 to 1,000 cal BP, and

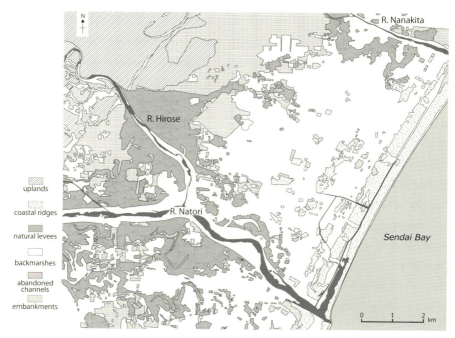

Fig. 2.2.16. Geomorphological map of northern part of Sendai Lowland
Landform classification is based on GSI Map (GSI HP)

around 1,000 cal BP, respectively. Coastal ridge I completed enclosure around 2,900 cal BP in the southern part of the lowland.

2.3. Beach Ridge Plains

2.3.1. Shimizu Lowland

The Shimizu lowland faces the western part of Suruga Bay (Figs. 1.1.1, 1.1.2; 2.3.1), and is surrounded by the Neogene mountains to the north and the Pleistocene Udo hills to the south. Within the lowland, the Tomoe River runs eastward and flows into Suruga Bay. Three coastal ridges, numbered I to III in a seaward direction, are remarkably developed in a north to south direction. Coastal ridges I and II may be recognized as Holocene marine terraces in front of the former sea cliffs to the east of Udo hill, which developed on the abrasion platforms. Furthermore, a sand and gravel spit, known as the Miho spit, is developed from the southeastern part of Udo hill towards the northeast (Figs. 2.3.2, 2.3.3). The sediments are considered to be supplied from both the Abe River, which flows in the west of the lowland, and the sea cliffs in the south to the southeastern part of Udo hill.

The ^{14}C date of a shell in the uppermost marine deposits of coastal ridge I at the eastern foot of Udo hill is around 6,600 cal BP (Matsushima, 1984 a). This suggests that the innermost coastal ridge at the eastern foot of Udo hill had already developed around the time of the culmination of the Holocene transgression. The Orido inlet exists between the coastal ridges and the Miho spit (Fig. 2.3.1), and a valley was formed on the surface of the Udo hill sediments beneath the Orido inlet, extending southward to a submarine canyon in Suruga Bay. Simultaneously, the Miho spit developed on a ridge on the surface of the Udo hill sediments. The recent deposits of the Shimizu lowland along the Tomoe River, from the bottom to the top, comprise fluvial sand, marine silt and clay, marine sand and gravel, silty sand and gravel with peat, and peaty silt and clay. The coastal ridges consist of marine sands and gravels that were deposited on top of the Udo hill sediments beneath the lowland (Figs. 2.3.4, 2.3.5, 2.3.6).

Sampling of the Holocene deposits was conducted in the backmarsh behind the Miho spit, at location S85, 2.2 m above mean sea level, and in the backmarsh between coastal ridges I and II, at location S95, 6.6 m above mean sea level (Fig. 2.3.1).

Although no ^{14}C date was obtained from the discontinuous bore-hole cores at S85, K-Ah volcanic glass dated to around 7,300 cal BP was found in the deposits at −21.3 m. Twenty-six samples were analyazed for foraminifera from the cores between −33.9 and −0.2 m, spanning an estimated period since ca. 8,600 cal BP. Fossil foraminifera were observed in the deposits be-

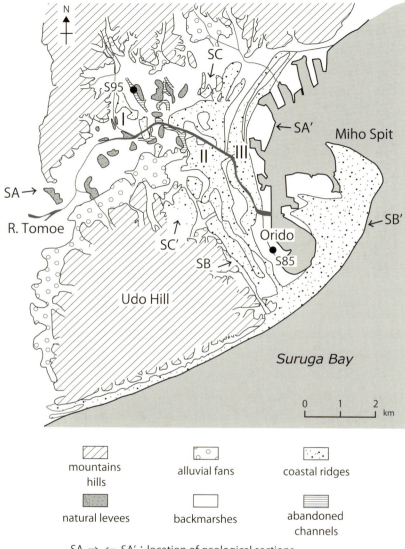

mountains hills	alluvial fans	coastal ridges
natural levees	backmarshes	abandoned channels

SA → ← SA' : location of geological sections

I ~ III : coastal ridges

S85, S95 : boring sites after Matsubara (2000a)

Fig. 2.3.1. Geomorphological map of Shimizu Lowland

Fig. 2.3.2. Miho Spit (Front. 12)
Mt. Fuji area including Miho Spit was inscribed in World Heritage List in 2013.

Fig. 2.3.3. Beach of Miho Spit

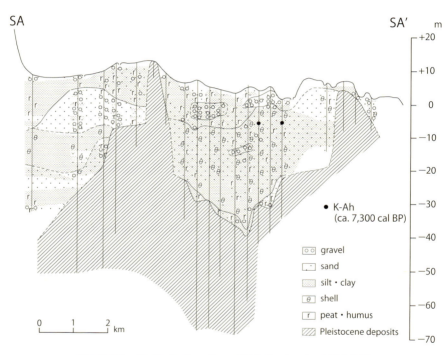

Fig. 2.3.4. Geological section of Shimizu Lowland (SA-SA')

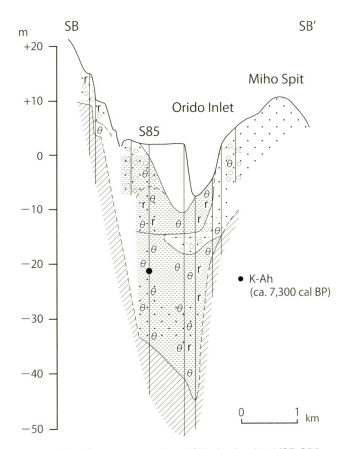

Fig. 2.3.5. Geological section of Shimizu Lowland (SB-SB')

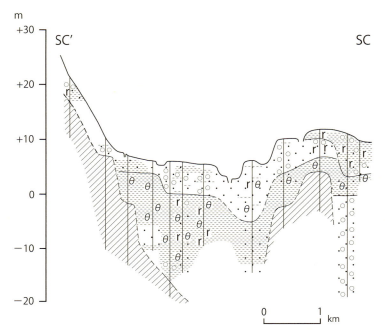

Fig. 2.3.6. Geological section of Shimizu Lowland (SC-SC')

tween −33.9 and −10.8 m (8,600 to 6,100 cal BP), and between −5.3 and −0.2 m (since ca. 5,500 cal BP).

The sediment may be divided into six parts (I to VI), according to their fossil foraminiferal assemblages as follows (Fig. 2.3.7).

I (−33.9 to −30.2 m, 8,600 to 8,200 cal BP)

Ammonia japonica, Elphidium advenum, and *E. jenseni* were dominant, accompanied by *A. beccarii* and *E. reticulosum*, which belong to the typical foraminiferal assemblage found in a bay. However, the frequency of the Indicator B species, suggesting the inflow of sea water from outside the bay, was still low. These results indicate that this period marked the beginning of the bay's expansion.

II (−30.2 to −23.2 m, 8,200 to 7,400 cal BP)

E. advenum, E. jenseni, A. japonica, and *Pseudononion japonicum* became

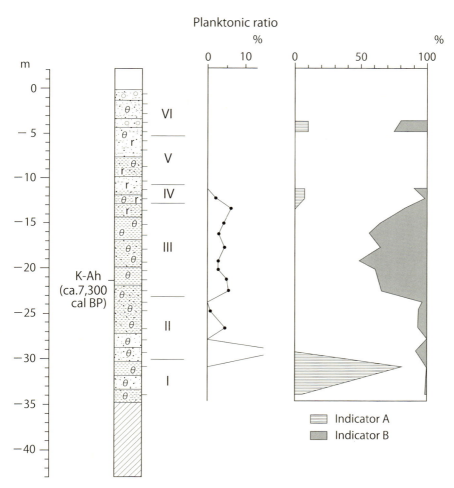

Fig. 2.3.7. Changes of fossil foraminiferal assemblages at location S85

dominant, while the occurrence of the Indicator B assemblage increased and planktonic foraminifera, indicating the inflow of open sea water, also occurred. Additionally, *Bulimina* cf. *fijiensis*, suggesting an environment with high water temperature, was also observed in these deposits.

III (−23.2 to −12.9 m, 7,400 to 6,400 cal BP)

The Indicator B assemblage accounted for almost 50 % of recovered fossils, whereas the proportion of planktonic foraminifera was around 5 %. In these deposits, *Bulimina* cf. *fijiensis* also occurred. These together suggest that both sea water from outside the bay and high-temperature open sea water flowed into the bay during this period.

IV (−12.9 to −10.8 m, 6,400 to 6,100 cal BP)

A. beccarii forma A, indicating a low salinity environment, were found alongside decreased frequencies of Indicator B and planktonic foraminiferal assemblages. Therefore, it is considered that the bay became enclosed during this period.

V (−10.8 to −5.3 m, 6,100 to 5,500 cal BP)

No foraminifera were found; therefore, it is inferred that only inland water flowed into the bay for this period.

VI (−5.3 to −0.2 m, since ca. 5,500 cal BP)

The number of fossils decreased, and planktonic foraminifera were not observed in these deposits, indicating that the influence of sea water was significantly reduced.

From these results, the geomorphic development around location S85 during the Holocene may be summarized as follows.

Sea water had already invaded the area behind the Miho spit through a narrow pass between Udo hill and the Miho spit around 8,600 cal BP. Between 7,400 and 6,400 cal BP, both sea water from outside the bay and high-temperature open sea water had a significant influence on the bay. However, the inflow of the sea water decreased after ca. 6,400 cal BP, and therefore it is considered that the Miho spit developed and began to connect with Udo hill to enclose the bay at this time.

At location S95, five ^{14}C dates were obtained from the continuous borehole cores, and K-Ah volcanic glass dated to around 7,300 cal BP was found in the deposits at −7 m. Seventeen samples from the cores were analyzed for foraminifera, between −15.9 and +0.5 m, spanning an estimated period from 8,500 to 4,200 cal BP.

The sediments may be divided into five parts (I to V), according to their fossil foraminiferal assemblages as follows (Fig. 2.3.8).

I (−15.9 to −15.1 m, before ca. 8,500 cal BP)

No foraminifera were found; therefore, it is considered that the sea water had not yet invaded the region.

II (−15.1 to −6.2 m, 8,500 to 7,100 cal BP)

A. beccarii forma B and C and *E. advenum* were dominant, accompanied by *Elphidium subgranulosum*, *A. japonica*, *E. jenseni*, and *E. reticulosum*, which belong to the typical foraminiferal assemblage found in a bay. The Indicator B species, suggesting the inflow of sea water from outside the bay, accounted for around 10%; furthermore, planktonic foraminifera, indicating the inflow of open sea water, also occurred. These together suggest that both sea water from outside the bay and open sea water had invaded the inner bay at this time.

III (−6.2 to −2.7 m, 7,100 to 6,600 cal BP)

The frequency of the Indicator B assemblage decreased and no planktonic foraminifera were found. This indicates that the influence of sea water from outside the bay was reduced.

IV (−2.7 to +0.4 m, 6,600 to 4,200 cal BP)

A. beccarii forma A, B, and C became dominant, whereas neither Indicator B nor planktonic foraminiferal assembalges were found. This indicates that the bay began to be enclosed by a coastal barrier and was transformed into a lagoon.

V (above +0.4 m, since ca. 4,200 cal BP)

No foraminifera were found in the peaty clay. Therefore, it is considered that the lagoon turned into a backmarsh after this time.

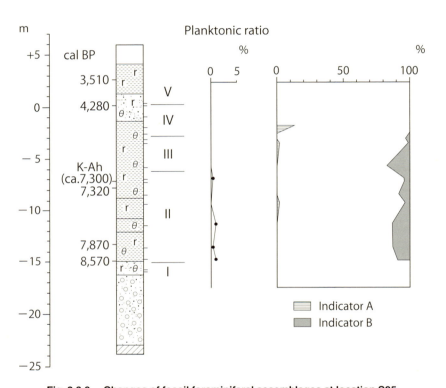

Fig. 2.3.8. Changes of fossil foraminiferal assemblages at location S95

From these results, the geomorphic development around location S95 during the Holocene may be summarized as follows.

Sea water began to invade the inner part of the lowland around 8,500 cal BP. Both sea water from outside the bay and open sea water had influence on the inner bay between 8,500 and 7,100 cal BP. The inflow of sea water, however, decreased after ca. 7,100 cal BP. As the inner bay behind coastal ridge II changed into a lagoon after ca. 6,600 cal BP, it is inferred that coastal ridge II began to emerge and enclose the bay at this time. Following this, coastal ridge II finished enclosing the lagoon around 4,200 cal BP. Coastal ridge III is considered to have been constructed between 1,800 and 1,700 cal BP, as indicated by the ^{14}C date of ca. 1,740 cal BP of the shell in the deposits just beneath the ridge sediments at −2.4 m (Matsushima, 1984 b).

According to the results obtained from both the S85 and S95 locations, the development of the coastal ridges can be summarized as follows (Fig. 2.3.9).

Sea water had already invaded the region by around 8,500 cal BP, and had also flowed through the narrow pass between Udo hill and the Miho spit. The innermost coastal ridge I had been constructed before at least ca. 6,600 cal BP, and the coastal ridge II began to develop seaward from ridge I and emerged to completely enclose the bay around 4,200 cal BP. Furthermore, the Miho spit developed and began to connect with Udo hill around 6,400 cal BP. Finally, the outermost coastal ridge III began to develop after 1,800 cal BP.

ca.7,000 cal BP

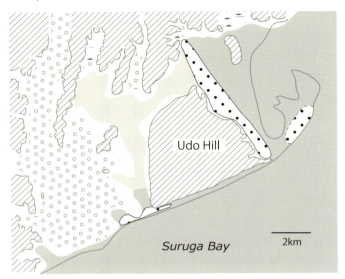

5,000 ~ 4,000 cal BP

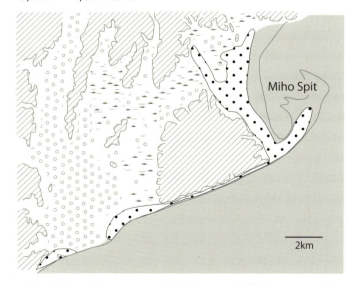

Fig. 2.3.9. Palaeogeographical changes in Shimizu Lowland

ca. 2,000 cal BP

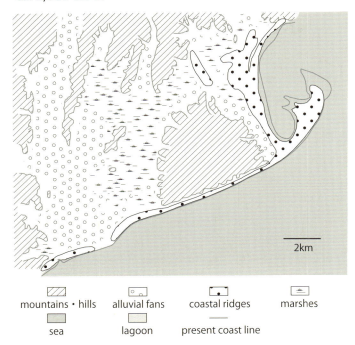

mountains • hills   alluvial fans   coastal ridges   marshes
sea   lagoon   present coast line

2.3.2. Tateyama Lowland

The Tateyama lowland is situated in the southwestern part of the Boso Peninsula, facing Tokyo Bay, and is surrounded by the Neogene hills. The Heguri and Shioiri Rivers rise in the hills and flow through the lowland into Tokyo Bay (Figs.1.1.1; 2.3.10, 2.3.11, 2.3.12, 2.3.13).

Four Holocene marine terraces are developed in the Tateyama lowland. These terraces have recorded the history of seismic crustal deformation as a result of large earthquakes along the Sagami Trough (Sugimura and Naruse, 1954, 1955; Yonekura, 1975; Yokota, 1978; Nakata *et al.* 1980; Kayane and Yoshikawa, 1986).

Six coastal ridges, numbered seaward from I to VI are developed in a north to south direction. Coastal ridges IV, V, and VI are particularly well

Chapter 2

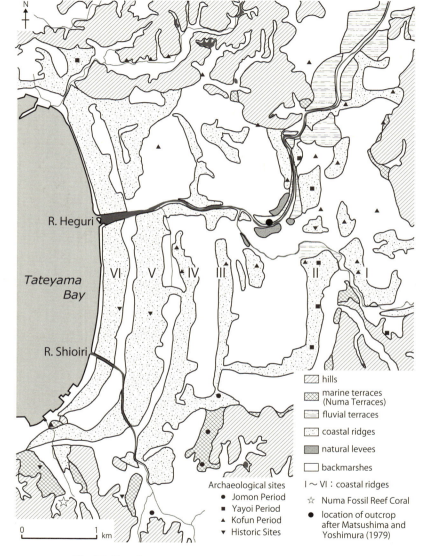

Fig. 2.3.10. Geomorphological map of Tateyama Lowland

Fig. 2.3.11. Tateyama Lowland from the southern hills

Fig. 2.3.12. Tateyama Bay

Fig. 2.3.13. Beach along Tateyama Bay

developed, and the borders between the ridges and the backmarshes are clear. The Tateyama lowland is composed of an alternation of sand and silt, with a maximum thickness of deposits of around 40 m along the present-day coast. The development of the coastal ridges is considered to have been caused by coseismic crustal movement associated with recurrence of two characteristic earthquakes: the Genroku (1703 AD) and Taisho (1923 AD) Kanto Earthquakes (Shishikura and Miyauchi, 2001).

Additionally, Fujiwara *et al.* (2006) examined the changes in depositional environment from shoreface to lagoon and from lagoon to inter-coastal ridge marsh, and attributed them to the coseismic uplifts accompanying the two historical earthquakes on the basis of ^{14}C dates. Coastal ridge VI had been formed by both a large-scale uplift of around 2.5 m by the Genroku Earthquake and a small-scale uplift of around 1.5 m by the Taisho Earthquake.

Fossil reef coral was discovered at Numa in the southwestern part of the lowland in the early 20th century, and was named the Numa fossil reef coral (Figs. 2.3.10, 2.3.14, 2.3.15). These fossils were estimated to date from 8,000 to 5,500 cal BP on the basis of ^{14}C dates by Yonekura (1975), Omoto (1976), and others. It has been inferred that the palaeotemperature of the sea water was around 4°C higher than that at present on the basis of the analysis of fossil assemblages by Hamada (1977). Furthermore, the palaeodepth of the sea was estimated to be 10 to 20 m, because fossil calcareous algae were not found. Fossil reef corals correlated with the Numa fossil reef coral were also observed in other locations of the Tateyama lowland, and the coastal lowlands of the Miura Peninsula and along Suruga Bay (Hamada, 1977).

Matsushima and Yoshimura (1979) analyzed the fossil molluscan assemblages in the Holocene marine deposits at a location behind coastal ridge III, approximately 2.5 km inland from the mouth of the Heguri River (Fig. 2.3.10). Oyster (*Hyotissa imbricate*) reefs were observed on the riverbed of the Neogene siltstone. 2.4 m-thick marine silt, and fluvial sand and gravel with a thickness of over 1.7 m were deposited on the oyster reefs (Figs. 2.3.16, 2.3.17). A trace fossil of a boring shell was found on the bedrock; therefore, the surface of the bedrock is inferred to have been a former wave-cut bench. [14]C dates of the boring shell indicate that the wave-cut bench was formed before ca. 7,600 cal BP. The fossil oysters, therefore, accompanied the fossil shells' inhabitants on reefs and fossil reef corals. The period of formation of these oyster reefs is estimated to be between 7,300 and 6,000 cal BP according to ^{14}C dates (Fig. 2.3.16).

Matsubara (2013) analyzed the fossil foraminiferal assemblages in the same marine deposits including the oyster reefs on the riverbed as those in which the fossil molluscan assemblages were analyzed. The sediments may be divided into three parts, numbered upward from I to III, according to their fossil foraminiferal assemblages as follows.

I (+2.5 to +3.5 m, 7,600 to 6,400 cal BP)

Pseudononion japonicum and *Pseudorotalia gaimardii*, which belong to the Indicator B assemblage and suggest the inflow of sea water from outside the

Fig. 2.3.14.　Numa Area

Fig. 2.3.15.　Numa Fossil Reef Coral preserved in Tateyama City

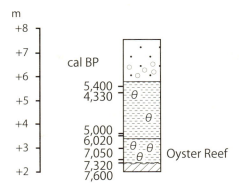

Fig. 2.3.16. Columnar section along Hegri River
The location of the section is shown in Fig.2.3.10.

bay, were dominant. *P. gaimardii* is known to only inhabit the area around the mouth of an embayment. These results suggest that sea water from outside the bay influenced the bay during this period.

Furthermore, *Cibicides lobatulus* adhering to rocks was found; *Elphidium crispum*, *E. jenseni*, and *E. reticulosum* adhering to seaweed on reefs were found; and *E. advenum* on the sandy sea bed in the middle to outer part of the embayment, was also found.

The occurrence of these species inhabiting reefs is influenced by the palaeoenvironment around the oyster reefs on the wave-cut bench. *Cibicides tenuimargo*, typical of tropical to subtropical shallow sea water, was observed in the deposits at around +3.0 m. This suggests that high-temperature sea water flowed into the bay around 7,000 cal BP.

II (+3.5 to +4.9 m, 6,400 to 4,800 cal BP)

Ammonia japonicum, inhabiting the middle to outer part of a bay, was dominant, whereas the Indicator B species, *P. japonicum* and *P. gaimardii*, decreased in the upper part of this section. These changes indicate that the influence of sea water from outside the bay gradually reduced during this period, which is in agreement with the analyses of fossil molluscan assemblages.

Fig. 2.3.17. Fossil Oyster Reef along Heguri River
The location of the outcrop is shown in Fig.2.3.10.

III (+4.9 to +5.8 m, 4,800 to 3,800 cal BP)

A. beccarii, which belongs to the typical foraminiferal assemblage found in a bay, increased in the deposits; however, no *Cibicides* spp. adhering to the rocks were found. This suggests a reduction in the bay. Fossil molluscan assemblages in this period further indicate that the bay became shallower. The reduction of the bay is inferred to have occurred as a result of filling by the deposits supplied from the rivers, in addition to the enclosure of the bay by coastal ridge III.

The geomorphic development in the Tateyama lowland according to the analyses of both fossil foraminiferal and fossil molluscan assemblages can be summarized as follows.

The bay began to expand by around 7,600 cal BP. Oyster reefs were formed on the wave-cut bench accompanied by reef coral between 7,300 and 6,000 cal BP. The bay was influenced by high-temperature open sea water between 7,600 and 6,400 cal BP; however, the inflow of sea water from outside the bay was reduced between 6,400 and 4,800 cal BP. Subsequently, the bay began to be filled with fluvial deposits and enclosed by coastal ridge III since around 4,800 to 3,800 cal BP.

Coastal ridges III to V are seen to correspond to Numa Marine Terrace III, in line with Nakata *et al.* (1980). Numa Terrace III was inferred to have emerged and completed enclosure of the bay around 2,880 cal BP. Therefore, the coastal ridges began to enclose the bay between 4,800 cal BP and 3,800 cal BP, and completed enclosure around 2,880 cal BP.

2.3.3. Kujukurihama Lowland

The Kujukurihama lowland is a typical beach ridge plain of Japan and situated in the eastern part of the Kanto Plain. It stretches from northeast to southwest, is approximately 60 km in length, and faces the Pacific Ocean (Figs. 1.1.1; 2.3.18, 2.3.19). As the lowland is located on the margin of the Kanto Basin, the crustal movements are characterized by uplift during the late Quaternary. The rivers flowing in the lowland include, from south to

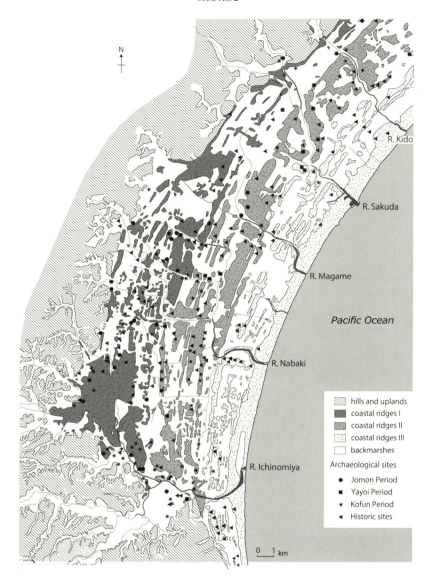

Fig. 2.3.18. Geomorphological map of Kujukurihama Lowland

Southwestern part (left), Northeastern part (right)

Geomorphic Development of Coastal Ridges in Japan

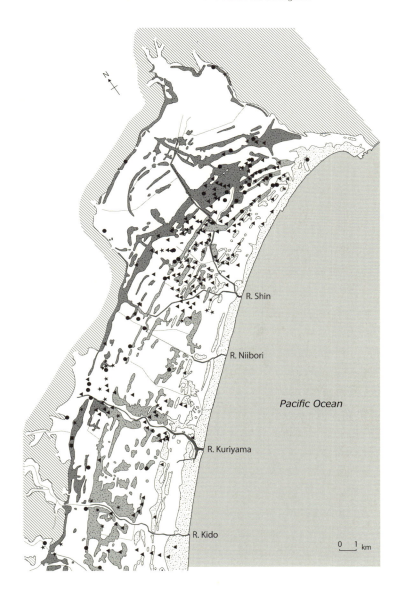

Chapter 2

Fig. 2.3.19.   Kujukurihama Lowland (Front. 2)

north, the Ichinomiya, Nabaki, Magame, Sakuda, Kido, Kuriyama, Shinbori, and Shin Rivers. All rivers intersect the coastal ridges and flow into the Pacific Ocean. The coastal ridges can be classified into three zones, numbered seaward from I to III, and are generally better developed in the southwestern part than in the northeastern part of the lowland.

Moriwaki (1979, 1982) examined the geomorphic development in the Kujukurihama lowland after the culmination of the Holocene transgression around 7,000 cal BP, and obtained the following results.

The coastal ridges had already developed between 6,500 and 6,000 cal BP. Completion of the formation of coastal ridges I, II, and III occurred during 5,500 to 4,500 cal BP, 4,500 to 2,500 cal BP, and around 1,500 cal BP, respectively.

Sampling of the Holocene deposits was conducted along the Magame River in the central part of the lowland by Masuda *et al.* (2001a; b) and Tamura *et al.* (2003). According to geological cross sections, a buried abra-

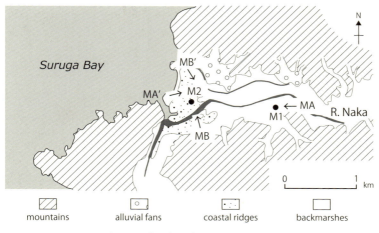

mountains alluvial fans coastal ridges backmarshes

MA → ← MA' : location of geological sections

• boring sites after Matsubara et al. (1986) and Matsubara (2000a)

Fig. 2.4.1.   Geomorphological map of Matsuzaki Lowland

sion platform occurs at approximately −10 m beneath coastal ridge I at around 7 km inland from the present-day coastline. This suggests that at least coastal ridge I developed on an abrasion platform during the Holocene.

2.4. Valley Plains

2.4.1. Matsuzaki Lowland

The Matsuzaki lowland faces Suruga Bay, and is situated in the lower reaches of the Naka River, which flows from the southern part of the central Izu Peninsula (Figs. 1.1.1, 1.1.2; 2.4.1). A coastal ridge encloses the mouth of the lowland, and the submarine topography off the coast shows a steep gradient.

The recent deposits in the Matsuzaki lowland, from the bottom to the top, comprise basal sand and gravel, lower silty sand with shell fragments,

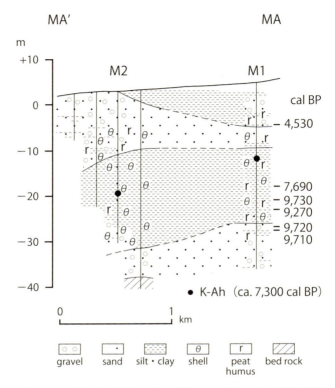

Fig. 2.4.2. Geological section of Matsuzaki Lowland (MA-MA')

lower sand and gravel, middle marine silt and clay, upper marine silty sand, and uppermost alluvial deposits (Figs. 2.4.2, 2.4.3).

Sampling of the Holocene deposits was conducted in the backmarsh behind the present-day coastal ridge (location M1, 5.5 m above mean sea level) and in the inner part of the coastal ridge (location M2, +3.0 m) (Fig. 2.4.1).

Almost undisturbed and continuous 41 m-long bore-hole cores were taken at M1. Six ^{14}C dates were obtained, and K-Ah volcanic glass dating to around 7,300 cal BP was found in the deposits at −12.1m. Thirty-nine samples from the cores between −24.1 and −4.9 m, spanning an estimated pe-

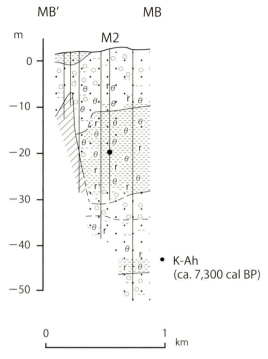

Fig. 2.4.3. Geological section of Matsuzaki Lowland (MB-MB')

riod from 9,400 to 5,100 cal BP, were analyzed for foraminifera. Fossil foraminifera were observed in the deposits between −21.8 and −5.9 m (9,000 to 5,600 cal BP).

The sediment may be divided into seven parts, numbered upward from I to VII, according to their fossil foraminiferal assemblages as follows (Fig. 2.4.4).

I (−24.1 to −21.8 m, 9,400 to 9,000 cal BP)

No foraminifera were found; therefore, it is inferred that the sea water had not yet invaded the region.

II (−21.8 to −20.7 m, 9,000 to 8,700 cal BP)

Elphidium jenseni, E. advenum, E. reticulosum, E. subgranulosum and *Ammo-*

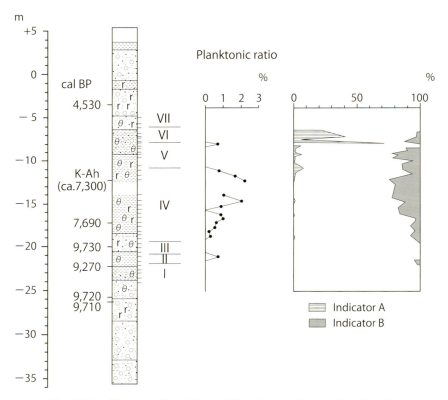

Fig. 2.4.4. Changes of fossil foraminiferal assemblages at location M1

nia beccarii forma B and C were found, which belong to the typical fossil foraminiferal assemblage found in a bay. The Indicator B species, suggesting the inflow of sea water from outside the bay, and planktonic foraminifera, indicating the inflow of open sea water, were also observed. These together suggest that the sea water began to invade the inner part of Matsuzaki lowland around 9,000 cal BP.

III (−20.7 to −19.3 m, 8,700 to 8,500 cal BP)

No foraminifera were found, and the sediments contain fluvial sand. Therefore, it is considered that only inland water flowed into the bay during this time.

IV (−19.3 to −10.8 m, 8,500 to 7,000 cal BP)

A. beccarii forma B and C, *Buliminella elegantissima, Elphidium somaense, E. subgranulosum, E. subincertum, E. advenum, E. jenseni, Pseudononion japonicum*, and *E. reticulosum* were found. Additionally, the frequencies of both Indicator B and planktonic foraminiferal assemblages were the greatest among all deposits. Furthermore, *Bulimina* cf. *fijiensis* was observed, suggesting an environment with high water temperature. These results indicate that the influence of sea water from outside the bay in addition to high-temperature open sea water reached a maximum during this period.

V (−10.8 to −7.7 m, 7,000 to 6,500 cal BP)

The frequency of *A. beccarii* forma A increased, suggesting a low salinity environment. Meanwhile, the number of planktonic foraminifera decreased. This suggests that the bay began to be enclosed by the coastal ridge.

VI (−7.7 to −5.9 m, 6,500 to 5,600 cal BP)

Although the frequency of *A. beccarii* forma A increased, the number of the Indicator B species decreased and no planktonic foraminifera were found. Consequently, the environment of the bay is considered to have acquired a considerably lower salinity and transformed into a lagoon.

VII (−5.9 to −4.9 m, 5,600 to 5,100 cal BP)

No foraminifera were found, and as the sediments contain fluvial sand

and gravel, it is considered that the lagoon began to be buried at this time.

Although no ^{14}C date was obtained from the discontinuous bore-hole cores of M2, K-Ah volcanic glass dated to around 7,300 cal BP was found in the deposits at −19 m. Eighteen samples were analyzed for foraminifera from the bore-hole cores between −28.5 and +1.9 m. Fossil foraminifera were observed in the deposits between −28.5 and −5 m, spanning an estimated period from 11,000 to 5,200 cal BP.

The sediments can be divided into seven parts, numbered upward from I to VII, according to their fossil foraminiferal assemblage as follows (Fig. 2.4.5).

I (−28.5 to −26.0 m, 11,000 to 9,700 cal BP)

E. advenum, E. jenseni and *P. japonicum* were dominant, accompanied by *A. beccarii* forma B and C. Furthermore, both Indicator B and planktonic foraminiferal assemblages occurred. These together suggest that the sea water began to invade the outer part of the Matsuzaki lowland after around 11,000 cal BP.

II (−26.0 to −21.1 m, 9,700 to 8,800 cal BP)

A. beccarii forma B and C, *E. somaense, E. subgranulosum, E. advenum, E. jenseni, P. japonicum*, and *E. reticulosum* were found, all of which belong to the typical foraminiferal assemblage found in a bay. The Indicator B species and planktonic foraminifera also occurred in addition to *Bulimina* cf. *fijiensis*, which suggests an environment with high water temperature.

III (−21.1 to −20.1 m, 8,800 to 8,600 cal BP)

No foraminifera were found, and as the sediments contain humus, it is considered that the inflow of the sea water decreased at this time.

IV (−20.1 to −16.5 m, 8,600 to 8,000 cal BP)

A. beccarii forma B and C, *E. subgranulosum, E. subincertum, E. somaense, E. advenum, P. japonicum*, and *E. reticulosum* were found. Furthermore, planktonic foraminifera and *Bulimina* cf. *fijiensis* were observed again. Consequently, it is inferred that the influence of high-temperature open sea water once again increased.

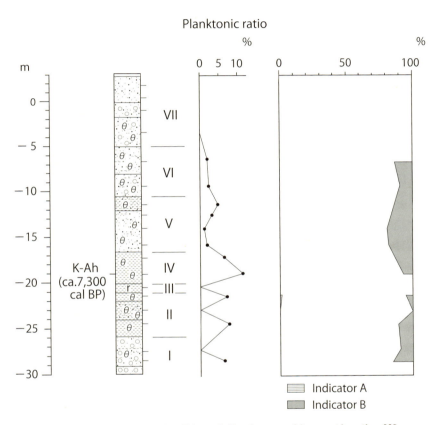

Fig. 2.4.5. Changes of fossil foraminiferal asssemblages at location M2

V (−16.5 to −10.5 m, 8,000 to 7,000 cal BP)

The benthic fossil foraminiferal assemblages were similar to that in IV. However, the frequency of planktonic foraminifera decreased, and *Bulimina* cf. *fijiensis* was not observed. This indicates that the bay began to be enclosed by the coastal ridge, which is supported by the fact that the deposits of this period are mainly composed of marine sand.

VI (−10.5 to −5.0 m, 7,000 to 5,200 cal BP)

The number of species of fossil foraminifera decreased. Additionally, the frequencies of both the Indicator B species and planktonic foraminiferal assemblages decreased. Consequently, the coastal ridge had been developing upwards since around 7,000 cal BP, and the bay began to be transformed into a lagoon.

VII (−5.0 to +1.9 m, after 5,200 cal BP)

No foraminifera were found; therefore, the coastal ridge had completely enclosed the lagoon at this time.

According to these results, the geomorphic development of the Matsuzaki lowland during the Holocene can be summarized as follows.

Sea water began to invade the region, causing a bay to be formed around 10,000 cal BP. The inflow of sea water from outside the bay and that of open sea water into the bay was greatest around 8,600 to 8,000 cal BP. During this period, the water temperature of the bay was higher than that at present. After around 7,000 cal BP, however, the environment of the bay acquired a low salinity condition, which is considered to be due to enclosure by a coastal ridge at the mouth of the bay. Enclosure by the coastal ridge changed the bay into a lagoon between 7,000 and 5,200 cal BP. Then, the lagoon was changed into a marsh following completion of the ridge's enclosure.

2.4.2 Haibara Lowland

The Haibara lowland faces the western part of Suruga Bay (Figs. 1.1.1, 1.1.2; 2.4.6, 2.4.7). It is surrounded by uplands, which consist of unconsoli-

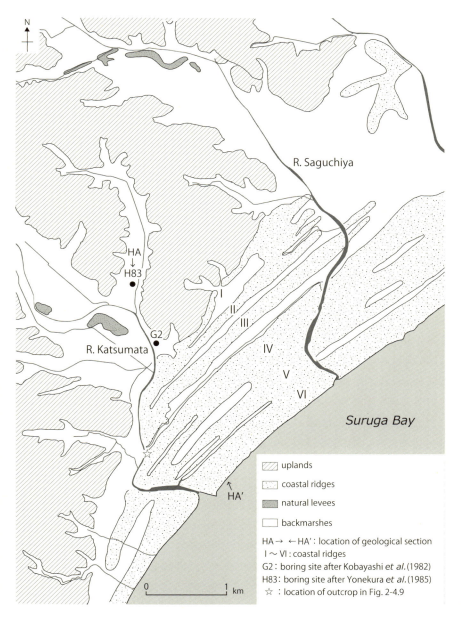

Fig. 2.4.6. Geomorphological map of Haibara Lowland

Fig. 2.4.7. Beach of Haibara Lowland

dated Pleistocene sediments, and it is situated near the mouth of the Oi River to the northeast. The Katsumata and Saguchiya Rivers in the lowland flow into Suruga Bay. Six coastal ridges, numbered I to VI in a seaward direction, are remarkably well developed on the buried abrasion platforms at approximately –5 m, in front of the former sea cliffs; they enclose the mouth of the valley plain. The sea bottom has a gentle gradient. Sand and gravel have been supplied to the lowland both from the Oi River and from the former sea cliffs.

The recent deposits of the Haibara lowland comprise, from the bottom to the top, an alternation of sandy silt and sand or gravel, silt and clay, humic silt and clay, and sand or gravel. The sandy silt contains shell fragments and humus. The coastal ridges consist of sand and gravel, which were deposited on top of silt and clay (Figs. 2.4.8, 2.4.9).

Holocene deposits were collected from undisturbed and continuous bore-hole cores at the locations H83, which is 4.2 m above mean sea level

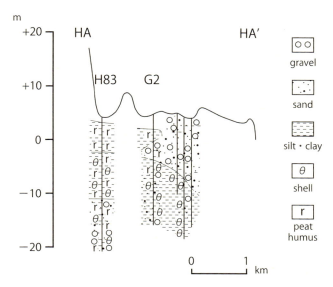

Fig. 2.4.8. Geological section of Haibara Lowland (HA-HA')

(Yonekura et al., 1985), and G2, which is 4.6 m above mean sea level (Kobayashi et al., 1982) (Fig. 2.4.6). Both locations are situated in the backmarsh behind the innermost coastal ridge I.

At location H83, 25 m-long cores were collected. Two ^{14}C dates were obtained, and K-Ah volcanic glass dated to around 7,300 cal BP was found at −6.3 m. Thirty-five samples from −19.5 to −1.9 m, spanning an estimated period from 8,900 to 6,700 cal BP were analyzed for foraminifera. Fossil foraminifera were observed in the deposits between −10.8 and −1.9 m (7,900 to 6,700 cal BP).

The sediment can be divided into four parts, numbered upward from I to IV, according to their fossil foraminiferal assemblages as follows (Fig. 2.4.10).

I (−19.5 to −10.8 m, 8.900 to 7.900 cal BP)

No foraminifera were found; therefore, it is considered that sea water had not yet invaded the region.

Fig. 2.4.9. Deposits of coastal ridge near R. Katsumata

The location is shown in Fig.2.4.6.

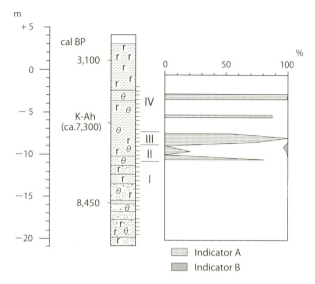

Fig. 2.4.10. Changes of fossil foraminiferal assemblages at location H83

II (−10.8 to −8.8 m, 7,900 to 7,600 cal BP)

Ammonia beccarii forma A, B, and C were dominant, accompanied by *Elphidium somaense, Buliminella elegantissima, E. advenum,* and *E. reticulosum.* The Indicator B species, suggesting the inflow of sea water from outside the bay, occurred in the deposits; however, planktonic foraminifera, which are indicators of the inflow of open sea water, were not found. Consequently, the influence of the sea water from outside the bay reached a maximum, but open sea water did not directly flow into the bay during this period.

III (−8.8 to −7.3 m, 7,600 to 7,400 cal BP)

A. beccarii forma A, B, and C were dominant, although the total number of fossil foraminifera species decreased. This suggests that the bay began to be enclosed by a coastal ridge at this time.

IV (−7.3 to −1.9 m, 7,400 to 6,700 cal BP)

Fossil foraminifera were observed in the deposits at −6.4 m, −5.4 m, −3.4 m, and −2.9 m. *A. beccarii* forma A, indicating a low salinity environment, was dominant. Thus, it is considered that, at this time, the bay was trans-

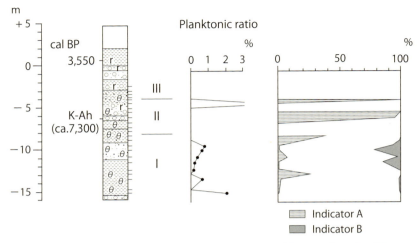

Fig. 2.4.11. Changes of fossil foraminiferal assemblages at location G2

formed into a lagoon.

After this, no fossil foraminifera were found in the deposits of humic silt and clay that began to accumulate above −1.9 m (since ca. 6,700 cal BP). Therefore, the lagoon is considered to have changed into a marsh around 6,700 cal BP.

At location G2, situated just behind coastal ridge I, 11 m-long cores were collected. One ^{14}C date was obtained, and K-Ah volcanic glass dated around 7,300 cal BP was found at −6.3 m. Twenty-three samples were analyzed for foraminifera from the bore-hole cores between −15.3 and −2.3 m. Fossil foraminifera were observed from −15.3 to −4.1 m, spanning an estimated period from 8,400 to 7,000 cal BP.

The sediment can be divided into three parts, numbered upward from I to III, according to their fossil foraminiferal assemblages as follows (Fig. 2.4.11).

I (−15.4 to −8.2 m, 8,400 to 7,500 cal BP)

A. beccarii forma B and C were dominant, accompanied by *A. beccarii* forma A, *B. elegantissima*, *E. somaense*, *E. subgranulosum*, *E. advenum*, *Pseu-*

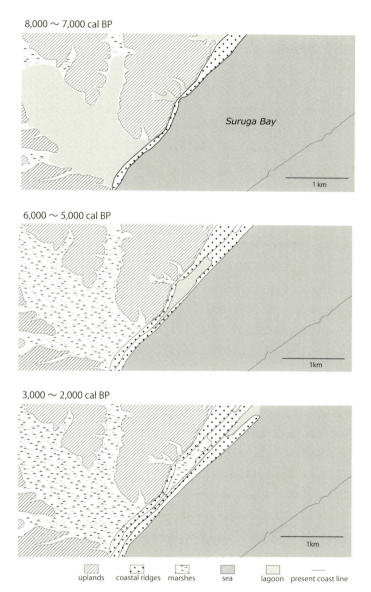

Fig. 2.4.12. Palaeogeographical changes in Haibara Lowland

dononion japonicum, and *E. reticulosum*, which belong to the typical foraminiferal assemblage found in a bay. Additionally, both the Indicator B species and planktonic foraminiferal assemblages occurred. Furthermore, *Bulimina* cf. *fijiensis* was observed in these deposits, suggesting an environment with high water temperature. Consequently, it is considered that both sea water from outside the bay and high-temperature open sea water flowed into the bay at this time. However, the occurrence of *A. beccarii* forma A suggests that the bay had begun to be enclosed by a coastal ridge.

II (−8.2 to −4.1 m, 7,500 to 7,000 cal BP)

Fossil foraminifera were observed in the deposits at −6.3 m, −5.6 m and −4.1 m. *A. beccarii* forma A was dominant; however, the Indicator B species did not occur. Consequently, the bay transformed into a lagoon during this period.

III (−4.1 to −2.3 m, since 7,000 cal BP)

No fossil foraminifera were found in the deposits of humic silt and clay. Therefore, the lagoon had begun to change into a marsh.

The geomorphic development in the Haibara lowland may be summarized according to the results obtained at both H83 and G2 as follows (Fig. 2.4.12).

A coastal ridge began to enclose the bay and transform it into a lagoon around 7,500 cal BP. Furthermore, the ridge completed enclosure of the lagoon to change it into a marsh between 7,000 and 6,500 cal BP. The periods of coastal ridges II and III formation are considered to be 6,000 to 5,000 cal BP, and 3,000 to 2,000 cal BP, respectively.

2.5. Delta–Beach Ridge Complexes

2.5.1. Sagami River Lowland

The Sagami River lowland faces Sagami Bay, and is surrounded by Pleistocene uplands (Fig. 1.1.1). A delta plain is recognized in the inner part of

GEOMORPHIC DEVELOPMENT OF COASTAL RIDGES IN JAPAN

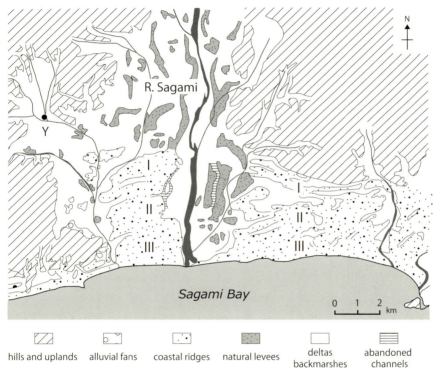

I ~ III : coastal ridges

Y : boring site after Matsuda *et al.* (1988)

Fig. 2.5.1.   Geomorphological map of Sagami River Lowland

Fig. 2.5.2. Eastern part of Sagami River Lowland

the lowland, alongside the Sagami River, whereas in the outer part of the lowland, three coastal ridges, numbered I to III in a seaward direction, are remarkably well developed (Figs. 2.5.1, 2.5.2). The coastal ridges in front of the eastern uplands are developed on buried abrasion platforms around 10 m below mean sea level.

The delta plain comprises, from the bottom to the top, basal gravel, lower and upper marine deposits, and uppermost fluvial deposits. The basal gravel is distributed in the buried valley of the former Sagami River, and the marine deposits are recognized approximately 11 km inland from the present-day coast.

Sampling of the Holocene deposits was conducted in the backmarsh behind the coastal ridges (location Y, at a height of 14.7 m above mean sea level) (Matsuda *et al.,* 1988) (Fig. 2.5.1). Seven ^{14}C dates were obtained from the continuous bore-hole cores. Twenty-two samples were analyzed for foraminifera between −13.2 and −3.6 m. Fossil foraminifera were observed in

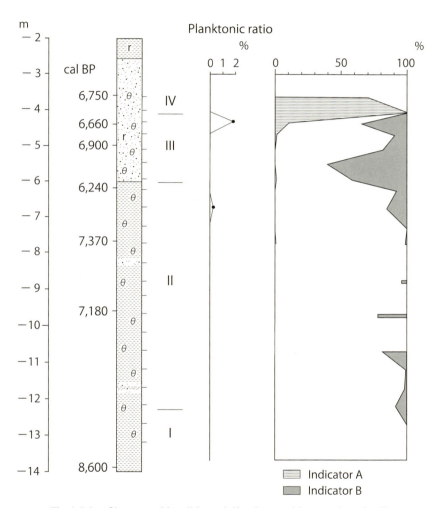

Fig. 2.5.3. Changes of fossil foraminiferal assemblages at location Y

the deposits between −12.5 and −3.6 m, spanning an estimated period from 8,200 to 6,000 cal BP.

The sediments may be divided into four parts, numbered upward from I to IV, according to their fossil foraminiferal assemblage as follows (Fig. 2.5.3).

I (−13.2 to −12.5 m, 8,400 to 8,200 cal BP)

No foraminifera were found; therefore, it is considered that the sea water had not yet invaded the area.

II (−12.5 to −6.1 m, 8,200 to 6,600 cal BP)

Elphidium somaense, E. subgranulosum, and *Buliminella elegantissima* were dominant, accompanied by *Ammonia beccarii* forma A, B, and C and *E. advenum*. The Indicator B species also occurred in the deposits, suggesting the inflow of sea water from outside the bay. Planktonic foraminifera, indicating the inflow of open sea water were observed in the deposits at −6.7 m. These results suggest that the bay was expanding during this period.

III (−6.1 to −4.2 m, 6,600 to 6,200 cal BP)

Pseudononion japonicum, one of the Indicator B species, and *A. beccarii* forma B and C were dominant. Furthermore, planktonic foraminifera were observed in the deposits at −4.3 m; the deposits consisted of sand. Consequently, the sea water is considered to have invaded and supplied sandy deposits into the bay from which the coastal ridges were constructed.

IV (−4.2 to −3.6 m, 6,200 to 6,000 cal BP)

A. beccarii forma A, indicating a low salinity environment, became dominant in the deposits. This suggests that the innermost coastal ridge I began to enclose the bay, transforming it into a lagoon.

As peaty clay was deposited above −2.2 m since ca. 5,600 cal BP, it is considered that the lagoon changed into a marsh after this time.

According to these results, the geomorphic development in the Sagami River lowland may be summarized as follows.

The bay began to be formed by around 8,200 cal BP and the inflow of sea water from outside the bay was greatest between 6,600 and 6,200 cal BP. During this period, coastal ridges were already beginning to be formed. As coastal ridge I started to enclose the bay around 6,200 cal BP, the bay was transformed into a lagoon. Following this, the lagoon changed into a marsh after around 5,600 cal BP, when the backmarsh deposits began to accumulate. Then, the outer coastal ridges II and III developed seaward from coastal ridge I. It is inferred that the outermost coastal ridge III began to be formed around 2,000 cal BP because of the ^{14}C date of 2,000 to 1,800 cal BP obtained from the sandy deposits of the ridge (Ota and Seto, 1968).

2.5.2. Obitsu River Lowland

The Obitsu River lowland faces Tokyo Bay, and is surrounded by the Pleistocene uplands (Fig. 1.1.1). The Obitsu River rises near Mt. Kiyosumi in the southern part of the Boso Peninsula and flows into Tokyo Bay. An arcuate delta develops in the coastal lowland. On the delta, both natural levees and coastal ridges develop (Figs. 2.5.4, 2.5.5). The Obitsu River delta, including tidal flats, is preserved as precious natural environment along Tokyo Bay.

Former sea cliffs are recognized at the northwest and the west of the uplands. The coastal ridges develop remarkably in front of the former sea cliffs. These coastal ridges develop on buried abrasion platforms aroud 5 to 10 m below mean sea level, in front of the former sea cliffs. The coastal ridges also develop at the foot of the northern uplands. According to the analyses of sedimentary facies of bore-hole data, Holocene marine deposits are recognized in the area around 10 km inland from the present-day coast. Consequently, it is inferred that these coastal ridges developed in front of the sea cliffs during the Holocene transgression.

A buried valley is recognized underneath the Holocene deposits. The recent formation in the lowland, from the bottom to the top, comprises fluvial sand, marine silt and clay, and alluvial deposits (Fig. 2.5.6)

Fossil foraminiferal assemblages are characterized by the typical foraminiferal assemblage found in a bay, such as *Ammonia beccarii, Elphidium subgranulosum*, and *E. incertum*.

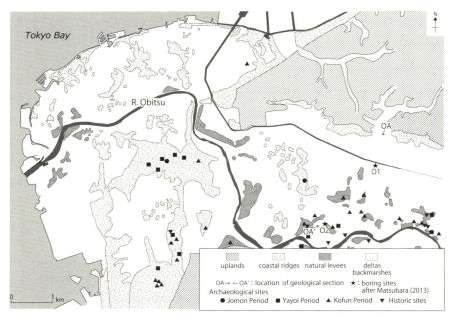

Fig. 2.5.4. Geomorphological map of Obitsu River Lowland

Fig. 2.5.5. Coastal ridges and natural levees along Obitsu River

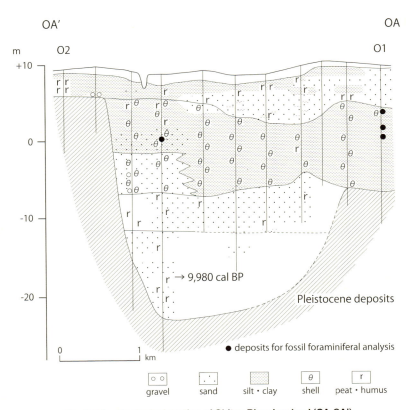

Fig. 2.5.6. Geological section of Obitsu River Lowland (OA-OA')

Chapter 3

Relationship between the Geomorphic Development of Coastal Ridges and Human Activities

3.1. Distribution of Archaeological Sites on Coastal Ridges

3.1.1. Lake Hamana and Hamamatsu Lowland

As described in Chapter 2, in Lake Hamana, the bay began to be formed around 10,500 cal BP, with sea water exerting the greatest influence on the bay between 9,000 and 8,000 cal BP. The barrier system began to develop during this period. When coastal ridge I, the former coastal barrier, began to enclose the bay around 7,500 cal BP, the bay was transformed into a lagoon. Furthermore, coastal ridges II and III began to develop seaward from coastal ridge I after around 5,000 cal BP.

The archaeological sites along Lake Hamana and in the Hamamatsu lowland are distributed mainly on the coastal ridges (Fig. 3.1.1).

Following a series of excavations at three archaeological sites, the Kajiko,

Kajiko-kita and Higashimae sites, in the Hamamatsu lowland, the relationship between the geomorphic development of the coastal ridges and human activities during the Holocene may be summarized as follows.

The Kajiko and Kajiko-kita sites are located in front of former sea cliffs in the eastern part of the lowland. The Kajiko site ("Kj" in Fig. 3.1.1) is situated on the outer part of coastal ridge I. Humans had advanced there between the middle Yayoi (ca. 2,000 cal BP) and the Heian (ca. 1,000 cal BP) periods. The 5 m-thick deposits on the surface may be divided into seven parts, numbered upward from I to VII (Matsubara, 2004). The deposits of coastal ridge I comprise sands, which are inferred to form the coastal ridges. On the basis of the ^{14}C date of ca. 6,960 cal BP of the peaty clay just above the deposits of coastal ridge I (Institute of Palaeoenvironmental Re-

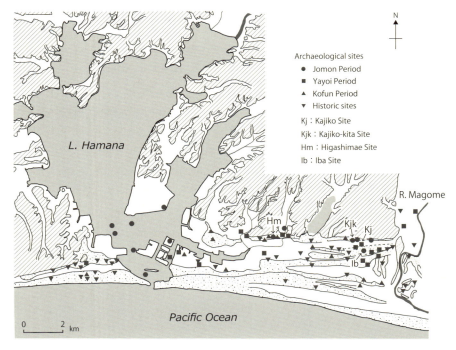

Fig. 3.1.1. Distribution of archaeological sites in Lake Hamana and Hamamatsu Lowland

search, 1994), it is inferred that the area around the Kajiko site on coastal ridge I began to be enclosed by the formation of coastal ridge II on the seaward side at this time. Consequently, coastal ridge I had completed its construction before 7,000 cal BP. Around the Kajiko site, buried abrasion platforms approximately 10 to 15 m below mean sea level are distributed in front of former sea cliff at the southern edge of the Pleistocene uplands. Therefore, it is supposed that the coastal ridges around the Kajiko site developed on these buried abrasion platforms.

The Kajiko-kita site ("Kjk" in Fig. 3.1.1) is also situated on the outer part of coastal ridge I. Humans advanced there between the early Jomon (7,000 to 5,500 cal BP) and the Heian (ca. 1,000 cal BP) periods. Excavations at the site showed that peaty clay was deposited on the sandy sediments. This peaty clay is considered to be deposited in backmarshes enclosed by coastal ridge II. A ^{14}C date of ca.3,400 cal BP was obtained from the bottom of the peaty clay. This suggests that the area around the Kajiko-kita site had been completely enclosed by around 3,400 cal BP. It is clear that coastal ridge I, on which the Kajiko-kita site is located, developed by at least ca. 7,000 cal BP, because humans began to settle there during the early Jomon period.

The Higashimae site ("Hm" in Fig. 3.1.1) is located on coastal ridge I, in front of former sea cliffs in the central part of the lowland. Humans first advanced there during the later Jomon period (4,500 to 3,000 cal BP).

Additionally, since the Jomon period, the Iba site ("Ib" in Fig. 3.1.1) is located on the eastern part of coastal ridge II. The Yayoi period moat of the Iba site was unearthed and preserved (Fig. 3.1.2).

CHAPTER 3

Fig. 3.1.2. Yayoi period moat of Iba Site in Hamamatsu Lowland

3.1.2. Kano River and Ukishimagahara Lowlands

As described in Chapter 2, in the Kano River and Ukishimagahara lowlands, the sea water invaded the region causing a bay to begin to form around 10,000 cal BP. Coastal ridge I, the former coastal barrier, began to form before 8,000 cal BP, emerging between 8,000 and 7,400 cal BP and completely enclosing the bay by around 6,800 cal BP. Subsequently, coastal ridge II enclosed the backmarsh around 5,000 cal BP, and coastal ridge III is considered to have been constructed before 2,000 cal BP.

On the delta that developed around the mouth of the Kano River, the Sanmaibashi and Numazu castles from the 16th and 18th centuries ("Sn" and "N" in Fig. 3.1.3), respectively, were built on coastal ridge II, which was constructed ca. 5,000 cal BP in the Ukishimagahara lowland. Archaeological excavations of the remains of the outer moat of Sanmaibashi castle have revealed remains dating to the latest Jomon period (ca. 3,000 BP). These suggest that humans began to advance on the ridge by at least 3,000 cal BP (Matsubara, 1998, 2000b). However, the coastal ridge was covered with vol-

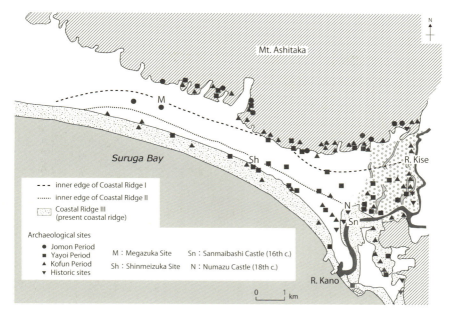

Fig. 3.1.3. Distribution of archaeological sites in Kano River and Ukishimagahara Lowlands

canic deposits from the Kawagodaira pumice fall (Kg) (Fig. 2.2.9) and pyroclastic flows from the Amagi volcano in the Izu Peninsula around 3,100 cal BP. Additionally, mudflows from the eastern side of Mt. Fuji affected the ridges. No archaeological remains have been found in these volcanic deposits. Human activity, such as the construction of the Sanmaibashi and Numazu castles, resumed around 500 BP on the Kano River delta. Therefore, it is clear that volcanic activity in this region strongly influenced human activity.

Although most of the archaeological sites in the Ukishimagahara lowland are distributed around the foot of Mt. Ashitaka and on the present-day coastal ridge III, the Megazuka archaeological site ("M" in Fig. 3.1.3) is located on coastal ridge I, the buried coastal barrier. This site dates between the middle Jomon (ca. 5,000 cal BP) and the later Kofun periods (ca. 1,500 cal BP). Following a series of excavations, the stratigraphic sequence and microlandforms at the Megazuka site, in addition to the periods of human activity, have been determined (Matsubara, 1992) (Figs. 3.1.4, 3.1.5, 3.1.6, 3.1.7, 3.1.8). During the initial stage of human activity in the middle to later Jomon period (5,000 to 4,000 BP), humans advanced onto the innermost coastal ridge I, setting up a campsite for fishing. Subsequently, humans began to settle on the coastal ridge during the latest Jomon period (ca. 3,000 cal BP). By the later Yayoi period (ca. 2,000 cal BP), permanent settlement had been maintained on the ridge for a long time. Excavation results confirm that humans were subsisting through both fishing and farming. After the early Kofun period (ca. 1,700 BP), however, human activity reduced, and human settlements at the Megazuka site were eventually abandoned during the later Kofun period (ca. 1,500 BP).

The initial stage of human activity coincided with the period in which the area around the Megazuka site changed from a lagoon to a marsh, with the construction of coastal ridge II around 5,000 BP. Coastal ridge I was not affected by sea water after the constructing of coastal ridge II. Additionally, the peak of human settlement coincided with the period during which the outermost coastal ridge III emerged and began to enclose the lagoon behind it around 2,000 cal BP.

Human settlement at the Megazuka site was abandoned at the time of the Obuchi Scoria fall (ObS) from Mt. Fuji (ca. 1,500 BP) (Fig. 2.2.9). Furthermore, the tectonic movements in the Ukishimagahara lowland are characterized by subsidence in the form of westward and landward downtilting, which buried the former coastal barrier (coastal ridge I) beneath the marsh. Continued burial of the ridge made the physical environment around the Megazuka site unsuitable for human settlement. These findings indicate that both volcanic activities and tectonic movements influenced human settlement in the Ukishimagahara lowland. However, many archaeological sites of the Kofun period are distributed on both coastal ridges II and III, such as the Shinmeizuka site ("Sh" in Fig. 3.1.3), located on coastal

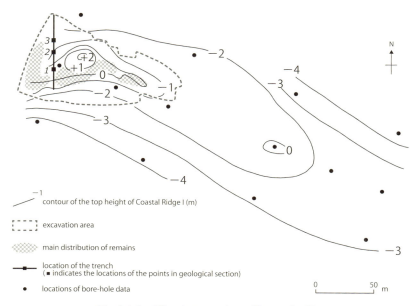

Fig. 3.1.4. Microtopography at Megazuka Site

Fig. 3.1.5. Marsh around Megazuka Site (Front. 8)

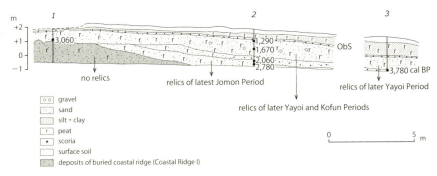

Fig. 3.1.6. Geological section along the trench
The location of the trench is shown in Fig.3.1.4.

Fig. 3.1.7. Surface deposits along the trench

Fig. 3.1.8. Remains of habitation during Yayoi period on buried coastal ridge at Megazuka Site (Front. 10)

ridge II. Earlier excavations at the Shinmeizuka site have revealed the fact that humans moved onto the coastal ridge II and subsisted through fishing from the middle Yayoi period (ca. 2,200 cal BP) (Matsubara, 2000b).

3.1.3. Shimizu Lowland

As described in Chapter 2, sea water began to invade the Shimizu lowland by around 8,500 cal BP. Subsequently, the sea water flowed through the narrow pass between Udo hill and the Miho spit. Coastal ridge I had been constructed prior to at least ca. 6,600 cal BP, following which coastal ridge II began to develop in a seaward direction and emerged to complete enclosure of the bay around 4,200 cal BP. The Miho spit developed and began to connect with Udo hill around 6,400 cal BP. Finally, coastal ridge III began to develop from around 1,800 cal BP.

The archaeological sites of both the Jomon and Yayoi periods in the Shimizu lowland are mainly distributed on coastal ridges I and II (Fig. 3.1.9). In particular, archaeological sites of the Jomon period can be found on coastal ridge I at the eastern foot of Udo hill. Human settlement began at the Tennouzan site ("T" in Fig. 3.1.9) on ridge I during the later to latest Jomon period (4,500 to 2,800 cal BP). Many archaeological sites of the Yayoi period (around 2,000 cal BP) are distributed on both coastal ridges I and II, and are also found on the natural levees along the Tomoe River. During the Yayoi period, humans settled at the Ishikawa II site ("I" in Fig. 3.1.9), situated on a natural levee. Additionally, archaeological sites of the Kofun period (ca. 1,500 cal BP) as well as more recent historic sites are distributed on coastal ridge III and on the Miho spit. Most significantly, human settlement has been identified at the Miyamichi site during the Kofun period ("M" in Fig. 3.1.9), located in the central part of the Miho spit. Excavations have shown that humans living there relied on fishing.

In the Shimizu lowland, during the later Jomon period (4,500 to 3,300 cal BP), humans began to advance on the innermost coastal ridge I, which developed to the east of Udo hill in around 7,000 to 6,500 cal BP. Since the Yayoi period (around 2,000 cal BP), humans have lived on both coastal ridge I in the central part of the lowland and on natural levees. Later, during the Kofun period (ca. 1,500 cal BP), human activity was observed and verified on coastal ridge II and the Miho spit, which were constructed between 5,000 and 4,000 cal BP. On the basis of these results, it is suggested that the innermost coastal ridge I, on which the Tennouzan site existed, was not sufficiently stable for human settlement until the outer coastal ridge II had fully formed around 5,000 to 4,000 cal BP. As the Miho island grew to border the eastern foot of Udo hill, the Miho spit also began to develop in around 6,500 to 6,000 cal BP. The strip of land at the root of the Miho spit, which was the former strait between the Miho island and Udo hill, was narrow; therefore, it is inferred that humans could not easily advance onto the Miho spit at this time. Instead, it took thousands of years for them to settle there.

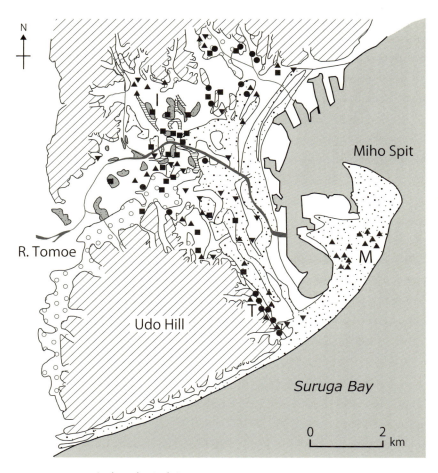

Archaeological sites

- Jomon Period
- Yayoi Period
- Kofun Period
- Historic sites

T : Tennouzan Site
I : Ishikawa Site II
M : Miyamichi Site

Fig. 3.1.9. Distribution of archaeological sites in Shimizu Lowland

3.1.4. Haibara Lowland

As described in Chapter 2, in the Haibara lowland, coastal ridge I began to enclose the bay and transform it into a lagoon around 7,500 cal BP, with the enclosure being almost complete between 7,000 and 6,500 cal BP, changing the lagoon into a marsh.

The archaeological sites in the Haibara lowland are mainly distributed on the inner coastal ridges I to III (Fig. 3.1.10). For example, the Shirayuri archaeological site ("S" in Fig. 3.1.10) is situated on coastal ridge I. Humans settled on coastal ridge I during the Yayoi period (ca. 2,000 cal BP) according to the results from a series of excavations at the Shirayuri site.

Humans have lived on coastal ridge I since the Yayoi period, despite it being constructed between 7,000 and 6,500 cal BP. Consequently, it may be supposed that it took thousands of years for coastal ridge I to become stabilized and free from the influence of sea water, possibly as a result of development of outer coastal ridges seaward from it.

3.1.5. Other Areas : Tateyama, Kujukurihama, Kuninaka and Obitsu River Lowlands

In the Tateyama lowland, archaeological sites from the Jomon period (before ca. 2,500 cal BP) are mainly distributed on the Holocene marine terraces—the Numa Marrine Terraces—in the southern part of the lowland, with few sites from this period occurring on the coastal ridges (Fig. 2.3.10). However, the archaeological sites dating to the Yayoi period (after ca. 2,500 cal BP) are mainly distributed on coastal ridge II. Additionally, sites from the Kofun period (after ca. 1,700 cal BP) are distributed on coastal ridges I to IV, particularly on I and II. A few historic sites are aslo located on coastal ridges V and VI. In the local context, coastal ridge I corresponds to Numa Terrace II, coastal ridges II to V correspond to Numa Terrace III, and coastal ridge VI corresponds to Numa Terrace IV (also known as the Genroku Terrace, uplifted during the Genroku Earthquake in 1703 AD).

Both the marine terraces and the coastal ridges developed as a result of

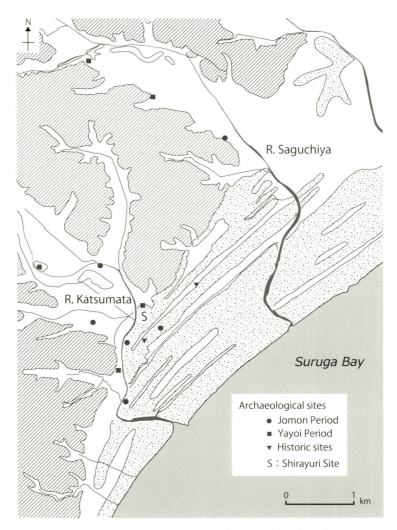

Fig. 3.1.10. Distribution of archaeological sites in Haibara Lowland

former seismic uplifts. It is inferred that the periods of emergence were around 4,200 cal BP for Terrace II and coastal ridge I, around 2,880 cal BP for Terrace III and coastal ridges II to V and around 250 cal BP for Terrace IV and coastal ridge VI (Nakata *et al.*, 1980). It is clear, therefore, that time lags of several hundred years can be recognized between the periods of coastal ridge formation and when humans began to advance and settle on the ridges.

The archaeological sites in the Kujukurihama lowland are mainly distributed on coastal ridges I and II, and most of the sites on coastal ridge III date to historical times (Fig. 2.3.18). The archaeological sites from the Jomon period are mainly distributed on coastal ridge I, which implies that the sites established during the Jomon period played a part in the fishery industry at that time (Kikuchi, 2001).

In the Kuninaka lowland, archaeological sites are distributed both at the foot of mountains and around coastal ridges (Fig. 2.2.14).

The archaeological sites in the Obitsu River lowland are mainly distributed on coastal ridges and natural levees (Fig. 2.5.4).

3.2. Distribution of Coastal Ridges and Human Activities in Tokyo

3.2.1. Geomorphic Development of the Kanto Plain

The Kanto Plain is surrounded by mountains to the north and west. The plain comprises hills, uplands, and lowlands. The hills were formed between 400,000 and 150,000 years ago. On the uplands, several terraces are developed, and the lowlands are distributed along the coasts of both Tokyo Bay and Sagami Bay and feature rivers such as the Tone, Ara, Tama, and Sagami Rivers (Fig. 3.2.1).

The sediment of the hills and uplands in the Kanto Plain is characterized by the deposition of Kanto loam, supplied mainly from the Hakone and Fuji volcanic mountains, located to the west of the plain. In the uplands,

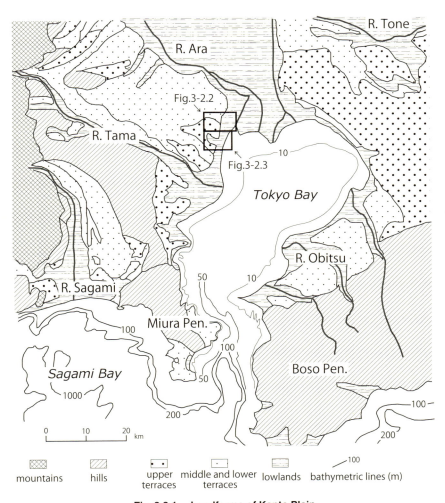

Fig. 3.2.1. Landforms of Kanto Plain

higher terraces are covered by older loam. The highest terrace in the uplands is known as Shimosueyoshi Terrace, and was formed during the Last Interglacial stage approximately 120,000 years ago. The lower terraces are, in order of age, Obaradai (ca. 100,000 years ago), Musashino (80,000 to 60,000 years ago) and Tachikawa (30,000 to 20,000 years ago). All these terraces are classified into three major groups: the upper (Shimosueyoshi and Obaradai), middle (Musashino), and lower (Tachikawa) terraces.

Palaeogeographical changes in the Kanto Plain since the Last Interglacial stage may be summarized as follows.

As a result of the transgression during the Last Interglacial stage, the Palaeo-Tokyo Bay expanded into the Kanto Plain. This transgression is locally called the Shimosueyoshi transgression, named after the Shimosueyoshi Terrace. During this period, marine sediments, which now comprise the higher terraces, were deposited. Later, the sea level lowered by around −120 m during the Last Glacial Maximum, and the land area of the Kanto Plain expanded offshore to include the area of present-day Tokyo Bay. During the Post-glacial stage, the Holocene transgression (locally the Jomon transgression) saw the sea invade a majority of the present lowlands in the Kanto Plain. After the peak of the transgression in around 7,000 cal BP, the present-day lowlands began to develop along the coasts and rivers.

3.2.2. Landforms of Central Areas in Tokyo

Landforms in urban areas have been changed artificially, including through reclamation of the foreshore and levelling of slopes. Natural landforms such as terraces can be recognized even in urban areas. However, some microforms such as abandoned channels cannot be found on the surfaces of present-day landforms because of the artificial changes to the lands. Recognizing the relationship between landform changes and human activities is important for urban planning including the prevention of disaster damage.

The Metropolis of Tokyo is situated in the southwestern part of the Kanto

Plain, facing Tokyo Bay. The Chiyoda, Chuo, and Minato wards comprise the main part of Tokyo, and Figs. 3.2.2 and 3.2.3, show the landforms and the distribution of archaeological sites in these three wards. The Imperial Palace (formerly Edo Castle, as Edo is the old name of Tokyo) is located on the upper terrace. Such terraces have been dissected, with valley plains developing arborescently upon them. Some moats around the castle were dug into these valley plains. The middle terraces are mainly developed along the Kanda River. In the coastal lowlands facing Tokyo Bay, a coastal ridge is developed from north to south, extending from the southern end of the Hongo Upland, belonging to the middle terraces (Fig. 3.2.2). A buried abrasion platform can be recognized beneath the coastal ridge, from isobaths of the bases of deposits laid down during the Post-glacial stage (Matsuda, 2009; Kaizuka, 2011). Consequently, it is considered that the coastal ridge developed on top of former abrasion platforms. Places such as Kanda, Nihonbashi and Ginza are located on the coastal ridge (Fig. 3.2.2). Additionally, valleys can be recognized beneath the lowland deposits to the west and east of the coastal ridge. The valley to the west of the ridge was once an inlet until the early 17th century, and the inlet and the coastal ridge were known as Hibiya-Irie (Hibiya Inlet) and Edo-Maejima (Edo Island), respectively (Figs. 3.2.4, 3.2.5). Another former abrasion platform can be recognized in the area between the Hongo Upland and the Sumida River, according to Matsuda (2009) and Kaizuka (2011). Holocene marine sediments are distributed along the Kanda River (Fig. 3.2.2).

The upper terraces are distributed around Minato ward to the south of Chiyoda and Chuo wards (Fig. 3.2.3), and valley plains are developed along the Shibuya-Furu River, dissecting the terraces. Coastal ridges are distributed about the eastern foot of the terraces, and are considered to have developed on buried abrasion platforms in front of the former sea cliffs. As Holocene marine deposits are seen along the Shibuya-Furu River around 2 km inland from the mouth of the river, it is inferred that these coastal ridges were formed by the Holocene transgression (Matsubara, 2012).

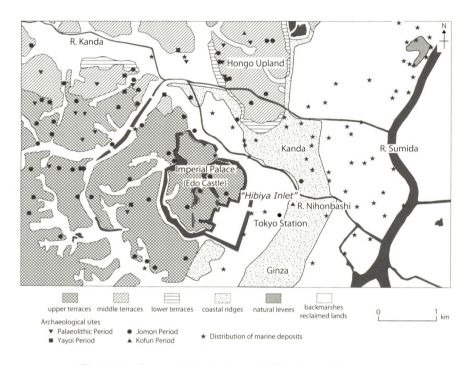

Fig. 3.2.2. Geomorphological map of Chiyoda and Chuo wards
The area in the Kanto Plain is shown in Fig.3.2.1. Landform classification is based on GSI Map (GSI HP).

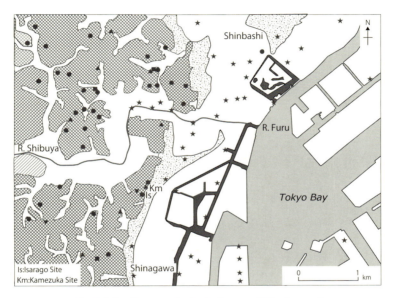

Fig. 3.2.3. Geomorphological map of Minato ward
The area in the Kanto Plain is shown in Fig.3.2.1. Landform classification is based on GSI Map (GSI HP).
Legends are the same as in Fig.3.2.2.

Fig. 3.2.4. Square in front of the Imperial Palace (former *Hibiya Inlet*)

Fig. 3.2.5. Imperial Palace on the upper terrace seeing from the square

3.2.3. Distribution of Archaeological Sites in the Centre of Tokyo

The archaeological sites are mainly found in the uplands, particularly at the edge of the terraces (Figs. 3.2.2, 3.2.3). Many shell mounds have been found at the eastern edge of the terraces facing the former Tokyo Bay (Fig. 3.2.6). The likely reason that these sites were distributed at the edge of the uplands is because the edge of the terraces represented a suitable place for fisheries, and water resources were available from springs from the terrace scarps (Figs. 3.2.7, 3.2.8). The archaeological sites in the lowlands are mainly distributed on the coastal ridges. This suggests that the coastal ridges that developed on the former abrasion platforms during the Holocene transgression had emerged earlier than the surrounding areas in the lowland to become stable and suitable for human settlement.

As mentioned above, human activities expanded mainly across the uplands during prehistoric times. However, landforms have been artificially changed during historical times, particularly during early-modern times (the Edo period). In particular, the shallows along Tokyo Bay have been reclaimed. This includes Hibiya Inlet, which underwent reclamation since

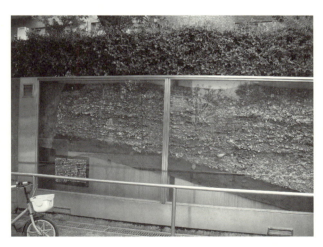

Fig. 3.2.6. Profile of Isarago shell mound in Minato ward
The location of the site is shown in Fig.3.2.3.

Fig. 3.2.7.　Terrace scarp (former sea cliff) near Kamezuka Site in Minato ward
The location of the site is shown in Fig.3.2.3.

Fig. 3.2.8.　Springs at the foot of terrace scarp near Kamezuka Site

1590 AD by Ieyasu Tokugawa, who established the Tokugawa shogunate in 1603 AD. Ieyasu Tokugawa began to reclaim the northern part of Hibiya Inlet using sediment supplied from the moats around Edo Castle since 1592 AD. Following this, he reclaimed the southern part of the inlet using sediment from the upland on the northern side of the castle (Endo, 2004; Matsuda, 2009). The purpose of this reclamation was to improve the living environment in Edo. During the Meiji era, dredging of Tokyo Bay and the mouth of the Sumida River was performed so that larger ships could use the harbours safely. Reclamation along the bay was then conducted using this dredged sludge. Finally, in the Showa period, particularly after WWII, the reclamation was undertaken on a larger scale because of increasing population and economic development (Koike and Ota, 1996; Masai, 2003; Endo, 2004).

3.3. Beach Erosion in Coastal Ridges

3.3.1. Inner Part of Suruga Bay: the Numazu-Fuji Coast

The Numazu-Fuji Coast is situated between the mouths of the Kano River to the east and the Fuji River to the west (Figs. 1.1.2; 2.2.1; 3.3.1, 3.3.2, 3.3.3). The lowland along the Numazu-Fuji Coast is recognized as a barrier–backmarsh system, as described in Chapter 2. The present-day coastal ridges consist of sand and gravel, supplied mainly from the Fuji River flowing along Fossa Magna, the plate boundary between the North American and Eurasian Plates in the central part of the Japanese Islands (Fig. 1.1.1). Furthermore, sand and gravel are sourced from the volcanic rocks distributed in the Kano River basin. Suruga Bay is situated at the plate boundary at which the Philippine Sea Plate subducts beneath the Eurasian Plate along the Suruga Trough (Fig. 1.1.2). The Suruga Trough is considered to be one of the hypocentral regions of massive earthquakes. Therefore, tsunamis are predicted to significantly influence the coast. Additionally, the

CHAPTER 3

Fig. 3.3.1. Numazu Coast

Fig. 3.3.2. Shingle beach of Numazu Coast (Front. 11)

Fig. 3.3.3. Fuji Coast

coast of the inner part of Suruga Bay often suffers from the damaging influence of high waves and high tides during typhoons, because high waves can reach the inner bay through the bay's wide mouth, which is open to the Pacific Ocean.

Coastal erosion, caused by a decrease in sediment supply from rivers, has become a serious problem since the 1960's. As a result, breakwaters have been constructed against coastal high waves and tsunamis. Additionally, the construction of offshore breakwaters and concrete armours has taken place, and beach nourishment has been implemented as a form of damage control against coastal erosion (Shibayama and Kayane, eds., 2013) (Figs. 3.3.1, 3.3.2, 3.3.3).

3.3.2. Miho Spit

As mentioned in Chapter 2 (Fig. 2.3.1), the Miho spit is located in the eastern part of the Shimizu lowland, and developed from the southeastern part of Udo hill towards the northeast (Figs. 3.3.4, 3.3.5). The sediments of the coastal ridges in the Shimizu lowland, including the Miho spit, are considered to have been supplied from the Abe River and the sea cliffs at the southern end of Udo hill during the Holocene transgression. The Abe River had supplied much sediment to develop the alluvial fan at the mouth of the river (Figs. 1.1.2; 3.3.6). The Abe River has been the main source of recent sediments.

A large amount of gravel was dug from the riverbed for construction materials in the period of economic growth after WWII in Japan; therefore, the sediments supplied to the Miho spit decreased rapidly, and severe coastal erosion began to occur along the spit. After gravel digging was prohibited in the late 1960s, sediment supply again increased in some parts of

Fig. 3.3.4.   Coast of Miho Spit (Front. 13)

Fig. 3.3.5. Shingle beach of Miho Spit

Fig. 3.3.6. Abe River Fan

Fig. 3.3.7. Beach nourishment of Miho Coast

Fig. 3.3.8. Offshore breakwaters of Miho Coast (Front. 14)

the coast. Later, offshore breakwaters were constructed and beach nourishment has been implemented to guard against coastal erosion (Uda, 2010) (Figs. 3.3.7, 3.3.8).

3.3.3. Amanohashidate

Concerning Amanohashidate (Fig. 2.1.9), coastal erosion was first recognized in this area during the Showa period after the sediment supply into Miyazu Bay for this coastal barrier from the rivers decreased because of river improvements such as the construction of sand-trap dams. Furthermore, two ports constructed to the north of the barrier have prevented sand supply from upcoast. Coastal dikes were constructed to protect against beach erosion after WWII. Particularly, since 1971, the larger dikes 30 m-long at intervals of 200 m began to be constructed. However, the decrease in sand supply could not be improved, and smooth coastlines along Miyazu Bay changed into serrated ones (Hirai, 1995). In an attempt to rectify this, sand bypass and recycling have been conducted since 1979, which have been effective means of beach nourishment (Shibayama and Kayane, eds., 2013)(Figs. 3.3.9, 3.3.10, 3.3.11).

CHAPTER 3

Fig. 3.3.9. Coastal changes in Amanohashidate from the southern mountains
[a: 1987 AD, b: 2014 AD] (Front. 5)

Lake Aso (left side), Miyazu Bay (right side)

Fig. 3.3.10. Present-day beach of Amanohashidate

Fig. 3.3.11. Effect of beach nourishment around coastal dikes

3.4. Influence of Tsunamis on Coastal Ridges :
Case of the 2011 Tohoku Earthquake

The Tohoku Earthquake was a massive earthquake (epicentre: 38°06.2' N, 142°51.6'E, Mw 9.0), which occurred on 11 March 2011. The source region of the earthquake was 450 km in length from north to south and 200 km in width from east to west along the Japan Trench, where the Pacific Plate is underthrusting westward beneath the North American Plate (Fig. 1.1.1). Tsunamis inundated wide areas along the coasts of the Tohoku and Kanto districts facing the Pacific Ocean.

The Sendai lowland faces the Pacific Ocean; it is around 50 km in length from north to south and 10 to 20 km wide from east to west. The Nanakita, Natori, Hirose, and Abukuma Rivers flow through the lowland into Sendai Bay. Most parts of the lowland have an altitude of less than 5 m, and the landforms in this area consist of natural levees along rivers and three coastal ridges (Figs. 2.2.16; 3.4.1).

At Sendai Bay, the maximum height of tsunami waves was recorded as more than 8.6 m at the Ishinomaki coast, which is located in the northern part of the bay (Japan Meteorological Agency: JMA, 2012). Shishikura et al. (2012) examined the distribution of tsunami deposits in the Sendai lowland. According to this study, in the northern part of the lowland, sandy deposits reached between 2.71 and 3.40 km inland, and muddy deposits reached 2.88 to 3.93 km inland. This is in contrast to the limit of inundation, which was 3.80 to 5.14 km inland (Geographical Survey Institute: GSI, 2014). In the central part of the lowland, the inland limits of sandy and muddy deposits were 3.14 km and 4.54 km, respectively. The limit of inundation was 5.05 km in this area. In the southern part of the lowland, sandy and muddy deposits were present up to 2.89 km and 3.05 km inland, respectively, whereas the limit of inundation was 3.72 km. These findings suggest that the distribution of tsunami deposits was smaller than the inundated area.

Coastal ridges I to III were all influenced by the tsunami (Haraguchi and Iwamatsu, 2011; GSI, 2014). However, the tsunami reached neither the natural levees along the Natori, Hirose and Nanakita Rivers nor the inner embankments in the backmarshes (Fig. 3.4.1). Abe *et al.* (1990) and Sawai *et al.* (2007) reconstructed the inundated areas of ancient tsunamis accompanying massive earthquakes in the Sendai lowland on the basis of analyses of historic documents, archaeological excavations and deposits in the lowland. According to these results, the tsunami from the Jogan Earthquake, which occurred in 869 AD, reached around 4 km inland, causing inundation of the backmarshes behind the coastal ridges.

The Ishinomaki lowland, which is a beach ridge plain, is located north-

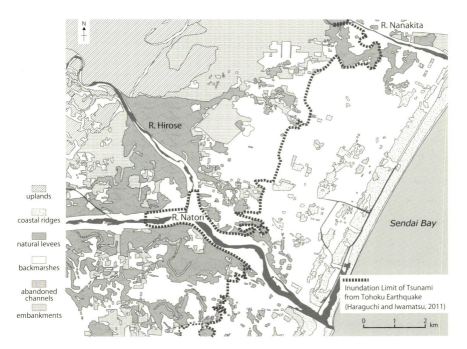

Fig. 3.4.1. Inundation area of tsunami from Tohoku Earthquake in the northern part of Sendai Lowland

Landform classification is based on GSI Map (GSI HP).

east to the Sendai lowland in the lower reaches of the Old Kitakami River, facing the northern part of Sendai Bay (No. 16 in Fig. 1.1.1, Tab. 1.1.1). The lowland is around 10 to 20 km in length from east to west and 50 km wide from north to south. Most parts of the Ishinomaki lowland have an elevation above mean sea level of less than 3 m. Five coastal ridges are distributed about 9 km inland from the present-day coastline (Matsumoto, 1984; Ito, 1999, 2003). The innermost coastal ridge began to develop during the Holocene transgression (Ito, 2003). In the western part of the Ishinomaki lowland, the inland limits of the distribution of sandy and muddy deposits were 1.87 km and 2.20 km, respectively (Shishikura *et al.*, 2012); however, the limit of inundation was 2.55 km inland (GSI, 2014). Although the outer three coastal ridges were influenced by the tsunami from the Tohoku Earthquake, the tsunami did not reach the areas around the inner two coastal ridges (Haraguchi and Iwamatsu, 2011; GSI, 2014).

As the Sendai and Ishinomaki lowlands are located along Sendai Bay, it is considered that the scale of the tsunami was large in both areas. However, the tsunami influenced the coastal ridges differently in these areas. Although all of coastal ridges I to III were influenced by the tsunami in the Sendai lowland, the tsunami did not reach the areas around the inner two coastal ridges in the Ishinomaki lowland. The difference was because the Sendai and Ishinomaki lowlands are different types of coastal lowlands. The landforms of the Sendai lowland are coastal ridge–backmarsh systems. However, in the Ishinomaki lowland, which is a beach ridge plain, coastal ridges are more strongly developed than in the Sendai lowland. Thus, it is inferred that the difference in the landforms of the coastal lowlands caused the differences in the inundation areas.

The tsunami from the Tohoku Earthquake also influenced the Kujukurihama lowland. The Kujukurihama lowland is a typical beach ridge plain in Japan, as described in Chapter 2 (Fig. 2.3.18). It is oriented northeast–southwest, and it is around 60 km in length facing the Pacific Ocean. The maximum of tsunami height in this region was recorded as 2.5 m at the

Choshi coast, which is situated at the northeastern end of the Kujukurihama lowland (JMA, 2012). Fujiwara *et al.* (2011, 2012) examined the distribution of the deposits carried by the tsunami from the Tohoku Earthquake, while Haraguchi and Iwamatsu (2011) and GSI (2014) showed the inundation areas (Fig. 3.4.2). Based on these studies, it is suggested that the tsunami reached the outermost coastal ridge III in the Kujukurihama lowland. Tsunami deposits were observed at locations behind the outermost coastal ridge, around 1 km inland from the present-day coastline in the central part of the lowland.

The influence of the tsunami from the Tohoku Earthquake on the landforms in three coastal lowlands, the Sendai, Ishinomaki, and Kujukurihama, was examined. The coastal ridges in each area, particularly the most seaward ridges, were influenced by the tsunami following the Tohoku Earthquake. All the coastal ridges in the Sendai lowland, which contains coastal ridge–backmarsh systems, were influenced by the tsunami. However, the tsunami did not reach the areas around the inner two coastal ridges in the Ishinomaki lowland, which is a beach ridge plain environment. In addition, the tsunami did not reach natural levees along the rivers and the inner embankments in the backmarshes of the Sendai lowland. These results suggest that the inundation area of tsunamis is influenced by small-scale landforms in coastal lowlands.

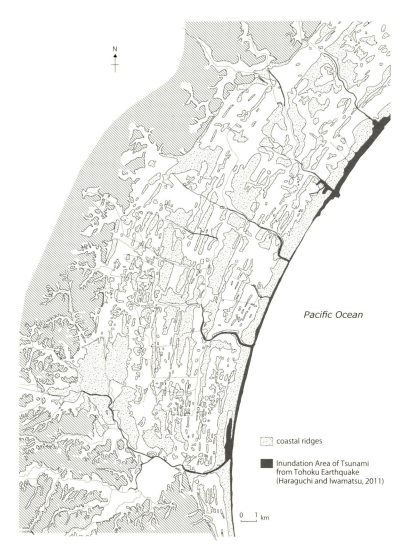

Fig. 3.4.2. Inundation area of tsunami from Tohoku Earthquake in Kujukurihama Lowland

Southwestern part (left), Northeastern part (right)

Relationship between the Geomorphic Developmemt of Coastal Ridges and Human Activities

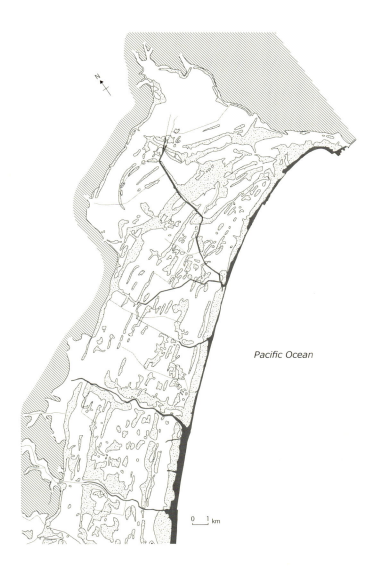

Pacific Ocean

Chapter 4
Geomorphic Development of Coastal Ridges during the Holocene

4.1. Common and Different Processes of Coastal Ridge Development

4.1.1. Common Processes in Relation to Relative Sea Level Change

A common trend in the relative sea level changes around Japan is that the sea rose above the present-day level (Ota *et al.*, 1981, 1982, 1990; Pirazolli, 1991; Bird, 2008). It is known that the sea level generally reached its highest level of 3 to 5 m higher than present at around 7,000 cal BP. After the culmination of the Holocene transgression, the sea level stabilized, or even lowered slightly, and has since undergone only minor fluctuations (Ota *et al.*, 1981, 1982, 1990; Umitsu, 1991). Fig. 4.1.1 shows the relative sea level curve in the Shimizu lowland.

Comparing the palaeoenvironmental changes in the studied areas, com-

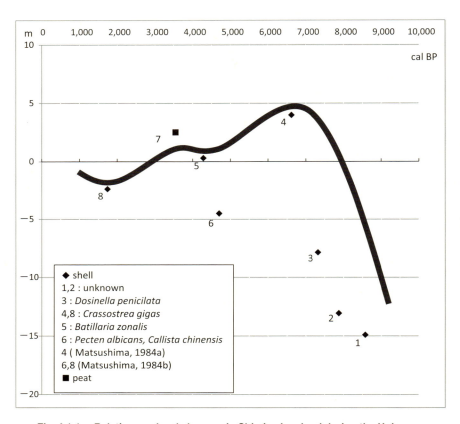

Fig. 4.1.1. Relative sea level changes in Shimizu Lowland during the Holocene

mon environmental changes are observed in the bay-expanding stage in each type of coastal lowland before around 8,000 to 7,000 cal BP, regardless of the regional differences in the antecedent topography and sediment supply. This implies that the relative sea level change was the dominant factor controlling the palaeoenvironment in a bay at least until 8,000 to 7,000 cal BP.

Furthermore, the processes of environmental change from bay to lagoon and to marsh or delta are common for each area. These palaeoenvironmental changes are related to coastal ridge development, and occurred in relation to the relative sea level changes.

The development of coastal ridges in each studied area can be summarized as follows (Fig. 4.1.2).

In Lake Hamana (a type: barrier–lagoon complexes), the Ukishimagahara and Tokoro lowlands (b type: sand and gravel ridge–backmarsh complexes), and the Haibara lowland (d-2 type: beach ridges in valley plains), the innermost coastal ridges began to enclose bays, and the bays changed into lagoons between 8,000 and 7,000 cal BP. In the Shimizu lowland (c type: beach ridge plains), the innermost coastal ridge began to be formed before around 6,500 cal BP. In the Hamamatsu lowland (c type), to the east of Lake Hamana, the period of formation for the innermost coastal ridge is estimated to be before 7,000 cal BP. Furthermore, in the Kuninaka lowland (b type) and the Matsuzaki lowland (d-1 type: barrier–backmarsh complexes in valley plains), the influence of enclosure by coastal ridges was seen after 7,000 cal BP. In the Sagami River lowland (e type: delta–beach ridge complexes), the innermost coastal ridges began to develop before 6,500 cal BP.

According to these results, the coastal ridges began to form during the Holocene transgression and started to enclose the bays without reference to the present-day landforms of the coastal lowlands. In some areas, the coastal ridges started to enclose the bays before the culmination of the Holocene transgression.

CHAPTER 4

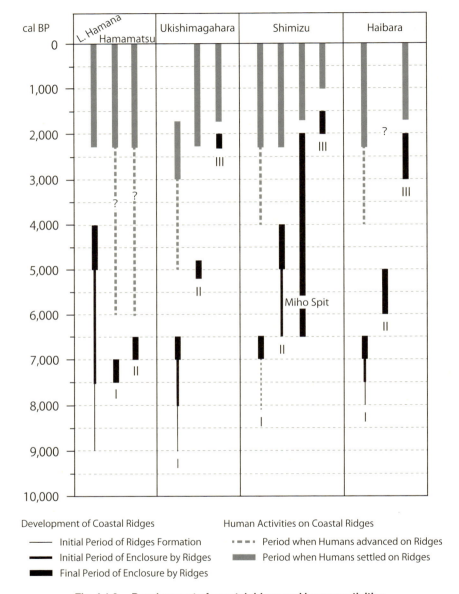

Fig. 4.1.2. **Development of coastal ridges and human activities**

Previous studies have implied that the coastal ridges began to form before 7,000 cal BP in Lake Kasumigaura (a type) (Saito *et al.*, 1990), Amanohashidate (a type) (Hirai, 1995), the Sarobetsu lowland (b type) (Sakaguchi *et al.*, 1985; Ohira, 1995), the Ishikari lowland (b type) (Sakaguchi, 1961, 1974; Matsushita, 1979), the Niigata lowland (b type) (Moriwaki, 1982), the Yufutsu lowland (c type) (Moriwaki, 1982), the Kujukurihama lowland (c type) (Moriwaki, 1979, 1982; Masuda *et al.*, 2001a; b), the Miyazaki lowland (c type) (Nagaoka *et al.*, 1991), and the Kimotsuki lowland (d-2 type) (Nagasako *et al.*, 1999).

Considering the influence of basal landforms on the development of coastal ridges, buried abrasion platforms were found to be distributed beneath the coastal ridges in the Hamamatsu, Kuninaka, Shimizu, Haibara, and Sagami River lowlands. It is considered that the coastal ridges, representing former coastal barriers in these areas, began to develop on the abrasion platforms during the Holocene transgression.

Common processes in the enclosure of the bay by coastal ridges can also be recognized. Typically, the innermost coastal ridges, representing former coastal barriers, began to enclose the bays, transforming them into lagoons between 8,000 and 7,000 cal BP. The basal deposits of the coastal ridges had already started to accumulate before ridges emerged to enclose the bays. Consequently, the rate of sea level rise is considered to be higher than the sedimentation rate before 8,000 cal BP. However, around 8,000 to 7,000 cal BP, the sedimentation rate exceeded the rate of sea level rise. The facies of the marine deposits showed no lithological changes in this period; therefore, the relationship between the sedimentation rate and the rate of sea level rise is not due to a change in the rate of sedimentation, but a change in the rate of sea level rise. Furthermore, the innermost coastal ridges typically completed enclosure of the lagoons, transforming them into marshes between 7,000 and 6,000 cal BP. This period corresponds with the time during which the relative sea level was stable or slightly lowered, just after the culmination of the Holocene transgression in Japan (Ota *et al.*, 1981, 1982,

1990; Pirazolli, 1991; Umitsu, 1991).

The period of development of coastal ridges II, constructed seaward from the innermost coastal ridges I in the Ukishimagahara and Shimizu lowlands corresponds to the first phase of regression after the culmination of the Holocene transgression around 5,000 cal BP (Ota *et al.*, 1981, 1982, 1990). Similar results have been obtained from previous studies on the Sarobetsu, Yufutsu, Kujukurihama, Miyazaki, and Kimotsuki lowlands. Coastal ridges III, developed seaward from coastal ridges II, in the Ukishimagahara, Shimizu, and Sagami River lowlands were constructed around 3,000 to 2,000 cal BP during the second phase of the regression (Ota *et al.*, 1981, 1982, 1990). Similar results have been obtained from studies on the Sarobetsu, Sendai, Kujukurihama, and Miyazaki lowlands.

Three stages in the development of coastal ridges in relation to relative sea level changes during the Holocene can be summarized as follows.

I. **Initial period of coastal ridge formation:** Before 8,000 cal BP, when the rate of rise in sea level was higher than the sedimentation rate, bays were formed by the Holocene transgression. At this stage, the innermost coastal ridges, representing former coastal barriers, had not yet emerged although the basal deposits in the ridges had begun to accumulate. The bays expanded as a result of the transgression, regardless of differences in the antecedent topography and sediment supply.

II. **Initial period of enclosure by coastal ridges:** Around 8,000 to 7,000 cal BP, the rate of sea level rise was decreasing; therefore, the sedimentation rate exceeded the rate of sea level rise. During this period, the innermost coastal ridges emerged and began to enclose the bays, and the bays were transformed into lagoons.

III. **Final period of enclosure by coastal ridges:** Around 7,000 to 6,000 cal BP, when the sea level was stable or slightly lowered, the innermost coastal ridges completed enclosure and transformed the lagoons into marshes. Then, the outer coastal ridges began to develop seaward from the inner ridges.

4.1.2. Factors Controlling Different Processes of Ridge Development 1 : Tectonic Movements

Case 1 : Ukishimagahara Lowland

The tectonic movements in the Ukishimagahara lowland are characterized by westward (towards the Suruga Trough) and landward downtilting, as discussed in Chapter 2. The rate of tectonic movement in this region during the Holocene was the highest along Suruga Bay. Even in such an active region, it is difficult to detect any influence of crustal movements on the development of the coastal lowlands until the period when the rate of sea level rise decreased. The influence of tectonic movements on the geomorphic development in the Ukishimagahara lowland can be recognized after around 8,000 to 7,000 cal BP.

In the lowland, the subsidence rate of the backmarsh side was higher than that of the coastal ridge side because of the landward downtilting. This suggests that the landward downtilting accelerated the enclosure of the bay by the coastal ridges, and that the periods in which the bay changed to a lagoon and then to a marsh were earlier than in the other study areas. Considering the present-day landforms, the Ukishimagahara lowland belongs to a barrier–backmarsh complex. However, there are buried coastal ridges distributed beneath the present backmarsh behind the present-day coastal ridge. It is, therefore, considered that the landward downtilting caused the former coastal ridges to subside.

The differences in the formation of coastal ridges can also be recognized in the Ukishimagahara lowland from east to west. The outer coastal ridges II and III in the eastern and central parts of the Ukishimagahara lowland have prograded seaward, whereas the ridges at the western end have developed upward. This is caused by local differences in the subsidence rate. The outer coastal ridges generally developed when the relative sea level was stable or slightly lowered. In the case of a lower subsidence rate, such as in the eastern and central parts of the Ukishimagahara lowland, the outer coastal ridges had grown seaward. In contrast, in the case of a more rapid subsidence rate, such as in the western part of the lowland, the outer ridges had grown upward in response to an apparent transgression.

Case 2 : Comparison between Ukishimagahara and Haibara Lowlands

In both the Ukishimagahara and Haibara lowlands, a similar geomorphic development of the coastal ridges can be reconstructed. In these areas, the innermost coastal ridges I, representing former coastal barriers, began to enclose the bays between 8,000 and 7,000 cal BP, and completed bay enclosure between 7,000 and 6,000 cal BP (Fig. 4.1.2). However, the formation of the coastal ridges differed in each area. In the Ukishimagahara lowland, situated in the east of the Suruga Trough, the innermost coastal ridge I and the outer coastal ridge II are recognized as buried landforms and deepen landward. In contrast, in the Haibara lowland, situated to the west of the Suruga Trough, all the coastal ridges (I to VI) can be recognized from landforms in the present coastal lowland. These differences are considered to be caused by regional differences in the characteristics of the tectonic movements: whether the area in question was subsiding or was relatively stable during the Holocene.

4.1.3. Factors Controlling Different Processes of Ridge Development 2 : Basal Landforms and Sediment Supply

Considering basal landforms and sediment supply, the geomorphic development in the Matsuzaki and Haibara lowlands will be compared. Although both lowlands belong to the valley plains, the periods of their palaeoenvironmental changes and the development of their coastal ridges do differ. It can be observed that the outer coastal ridges developed seaward from the innermost coastal ridge in the Haibara lowland. However, such seaward development of ridges is not apparent in the Matsuzaki lowland, and these differences are considered to be caused by the differences in basal topography and sediment supply.

In the Haibara lowland, buried abrasion platforms were distributed beneath the coastal lowland, whereas in the Matsuzaki lowland, there is no evidence of such landforms. Moreover, in the Hibara lowland, abundant sediments are supplied from a sea cliff and a large river (the Oi River), whereas the Matsuzaki lowland does not have a significant source of sediments. Therefore, the presence of basal landforms in the coastal lowlands and the abundant sediment supply can be observed to accelerate the development of coastal ridges in the Haibara lowland. These differences further resulted in the earlier enclosure of the bay by the innermost coastal ridge in the Haibara lowland.

4.2. Human Settlement on Coastal Ridges during the Holocene

Human settlement on the coastal ridges in four study areas can be summarized as follows (Fig. 4.1.2).

Lake Hamana and the Hamamatsu lowland belong to barrier-lagoon complexes and beach ridge plains, respectively. The inner ridges, among the six coastal ridges in the Hamamatsu lowland, developed on buried abrasion platforms. The archaeological sites around Lake Hamana and in the Hamamatsu lowland are distributed on the coastal ridges themselves. In particular, the sites mainly occur on the inner coastal ridges in the Hamamatsu lowland. The basal deposits of these coastal ridges began to accumulate around 9,000 cal BP, and the innermost coastal ridge I began to enclose the bay, transforming it into a lagoon between 8,000 and 7,000 cal BP. At this stage, humans had not yet advanced onto coastal ridge I. Between 5,000 and 4,000 cal BP, the formation of coastal ridge I was completed, and outer coastal ridges II and III developed seaward. At this stage, humans began to advance onto coastal ridge I. Around 2,000 cal BP, humans began to settle on coastal ridges I and II.

The Ukishimagahara lowland contains one coastal ridge and a backmarsh area. However, two former coastal ridges are buried behind the present-day ridge, because the lowland has been downtilted landward. Most of the archaeological sites in the Ukishimagahara lowland are distributed at the foot of the mountains and on the present-day coastal ridge. However, some archaeological sites do exist on the buried coastal ridge I. Megazuka is a habitation site located on the buried coastal ridge I. Humans began to advance onto the ridge after around 5,000 cal BP, and settled there between 3,000 and 2,000 cal BP. In the Ukishimagahara lowland, the innermost coastal ridge I started to develop around 9,000 cal BP. The ridge began to enclose the bay between 8,000 and 7,000 cal BP, changing the bay into a lagoon. At this stage, humans had not yet advanced onto the ridge. Subsequently, coastal ridge I emerged and completed enclosure between 7,000 and 6,000 cal BP. Around 5,000 cal BP, the outer coastal ridge II developed seaward from the ridge I. At this stage, humans began to advance onto coastal ridge I. Around 3,000 to 2,000 cal BP, when the outermost coastal ridge III developed, humans had begun to settle on coastal ridge I. Later, tectonic movements in the form of westward and landward downtilting, and volcanic activity, in the form of scoria fall from Mt. Fuji caused the abandonment of the settlement on coastal ridge I around 1,500 cal BP. Furthermore, in the Kano River lowland in the east of the Ukishimagahara lowland, humans had advanced on the coastal ridges since 3,000 cal BP. However, volcanic activities such as pumice fall and pyroclastic flows from the Amagi volcano and mudflows from Mt. Fuji disrupted their establishment on the ridge.

The Shimizu lowland belongs to the beach ridge plains group. Three coastal ridges and the Miho spit developed in the lowland. The archaeological sites during both the Jomon and Yayoi periods in the Shimizu lowland are distributed on coastal ridges I and II, whereas on the Miho spit, only the sites since the Kofun period were recognized. Human settlement extended seaward along the development of the coastal ridges. Humans began to advance and settle on coastal ridge I during the period in which coastal ridges II and III, and the Miho spit were developing seaward. Between 7,000 and 6,000 cal BP, the formation of coastal ridge I was completed, and coastal ridge II developed seaward. Additionally, the Miho spit began to form, connecting with Udo hill at this time. Around 2,000 cal BP, when coastal ridge III was developing, humans settled on coastal ridges I and II.

The Haibara lowland belongs to the valley plains. Six coastal ridges are developed in the lowland. These are characterized by their development on buried abrasion platforms. The archaeological sites in the Haibara lowland are mainly distributed on the inner coastal ridges. Between 8,000 and 7,000 cal BP, the innermost coastal ridge I began to enclose the bay, transforming it into a lagoon. At this stage, humans had not yet advanced onto the ridge. Between 7,000 and 6,000 cal BP, the formation of coastal ridge I was completed. During this period, human settlement on the coastal ridges was still not observed. Humans began to advance onto coastal ridge I after around 4,000 cal BP. During 3,000 to 2,000 cal BP, humans began to settle on coastal ridge I.

According to these results in the four study areas, the coastal ridges in the lowlands may be considered important landforms for human settlement during the Holocene. Human activity has been influenced by the development of coastal ridges and devastating volcanic events. Time lags are recognized between the final stages of coastal ridge formation and the stage when humans began to advance and settle on the ridges. These time lags represent the fact that it took thousands of years for the coastal ridges to become stabilized and free from the influence of sea water, including high waves, high tides, and tsunamis. This is supported by the fact that coastal ridges, particularly the outermost ridges in the coastal lowlands along the Pacific coast, were influenced by the tsunami following the Tohoku Earthquake.

References

Abe, H., Sugeno, Y. and Chigama, A. 1990: Estimation of the height of the Sanriku Jogan 11 Earthquake-Tsunami (A. D. 869) in the Sendai Plain.*Earthquakes*, 43, 513–525. (J+E)

Akagi, S., Toyoshima. Y., Hoshimi, K. and Tanimura, M. 1993: Geoenvironment and geologic history of Lake Koyamaike, Tottori Prefecture. *Mem. Geol. Soc. Japan*, 39, 103–116. (J+E)

Akamatsu, M., Kitagawa, Y., Matsushita, K. and Igarashi, Y. 1981: ^{14}C-age of the shell beds in Sarobetsu moor and Ishikari coastal plain, Hokkaido. *Chikyu Kagaku (Earth Science)*, 35, 215–218. (J)

Ariga, T. 1984: Geomorphological development of Shonai coastal plain, northeastern Japan: The behavior of alluvial fan's front since latest Pleistocene. *Annals, Tohoku Geogr. Assoc.*, 36, 13–24. (J+E)

Beets, D. J., van der Valk, L. and Stive, M. J. F. 1992: Holocene evolution of the coast of Holland. *Mar. Geol.*, 103, 423–443.

Bird, E. C. F. 2008: *Coastal Geomorphology —An Introduction*. 2nd edition. (Wiley), 411p.

Bird, E. C. F. and Jones, D. J. B. 1988: The origin of foredunes of the coast of Victoria, Australia. *Jour. Coastal Res.*, 4, 181–192.

Chiji, M. and Lopez, S. M. 1968: Regional foraminiferal assemblages in Tanabe Bay, Kii Pen., Central Japan. *Pub. Seto Mar. Biol. Lab.*, 16, 85–125.

Colquhoun, D. J., Pierce, J. W. and Schwartz, M. L. 1968: Field and laboratory observations on the genesis of barrier island. *Geol. Soc. Amer. Ann. Meeting. Abstr.*, 59–60.

Cushman, J. A. 1942: The formation of the tropical Pacific collections of the "Albatross", 1899–1900. *Smithsonian Inst. U. S. Nat. Mus., Bull.*, 161, 139p.

Davis, R. A. 1994a: Barrier island systems —a geologic overview—. In Davis, R. A. (ed.), *Geology of Holocene Barrier Island Systems* (Springer-Verlag), 1–46.

Davis, R. A. 1994b: Other barrier systems of the world. In Davis, R. A. (ed.), *Geology of Holocene Barrier Island Systems*. (Springer-Verlag), 435–456.

Dillenburg, S. and Hesp, P. (ed.) 2009: *Geology and Geomorphology of Holocene Coastal Barriers of Brazil.* Lecture Notes in Earth Sciences, 107, (Springer), 380p.

Endo, T. 2004: Historical review of reclamation works in the Tokyo Bay Area. *Jour. Geogr.*, 113, 785–801. (J+E)

Fuji, N. 1975: The coastal sand dunes of Hokuriku district, Central Japan. *The Quat. Res.*, 14, 195–220. (J+E)

Fujimoto, K. 1988: Beach-ridge ranges and their formative periods on three Holocene coastal plains in the southeastern part of Fukushima Prefecture, Northeastern Japan. *Annals, Tohoku Geogr. Assoc.*, 40, 139–149. (J+E)

Fujiwara, O., Hirakawa, K., Irizuki, T., Kamataki, T., Uchida, J., Abe, K., Hasegawa, S., Takada, K. and Haraguchi, T. 2006: Progradation of Tateyama strand plain system, SW coast of Boso Peninsula, Central Japan, triggered by coseisomic uplifts during the historical Kanto Earthquakes. *The Quat. Res.*, 45, 235–247.(J+E)

Fujiwara, O., Sawai, Y., Morita, Y., Komatsubara, J. and Abe, K. 2007: Coseismic subsidence recorded in the Holocene sequence in the Ukishima-ga-hara lowland, Shizuoka Prefecture, central Japan. *Annual Report on Active Fault and Paleoearthquake Researches No. 7* (AIST Geological Survey of Japan), 91–118. (J+E)

Fujiwara, O., Irizuki, T., Sampei, Y., Haruki, A., Tomotsuka, A. and Abe, K. 2008: Holocene environmental changes in the Ukishima-ga-hara lowland, along the Suruga Bay, central Japan. *Annual Report on Active Fault and Paleoearthquake Researches No. 8* (AIST Geological Survey of Japan), 163–185. (J+E)

Fujiwara, O., Sawai, Y., Shishikura, M., Namegaya, Y. and Kagohara, K. 2011: Sedimentary features of the 2011 Tohoku earthquake tsunami deposit, on the Hasunuma coast (central part of Kujukuri coast), east Japan. *Annual Report on Active Fault and Paleoearthquake Researches No. 11* (AIST Geological Survey of Japan), 97–106. (J+E)

Fujiwara, O., Sawai, Y., Shishikura, M. and Namegaya, Y. 2012: Sedimentary features of the 2011 Tohoku earthquake tsunami deposits on the central Kujukuri coast, east Japan. *The Quat. Res.*, 51, 117–126. (J+E)

Fujiwara, O., Sato, Y., Ono, E. and Umitsu, M. 2013: Researches on tsunami deposits using sediment cores: 3.4 ka tsunami deposit in the Rokken-gawa lowland near Lake Hamana, Pacific coast of central Japan. *Jour. Geogr.*, 122, 308–322. (J+E)

Geographical Survey Institute (GSI) : http://www.gsi.go.jp/

GSI 2014: Maps of inundation areas by Tohoku Earthquake (1: 25,000). www.gsi.go.jp/kikaku/kikaku40014.html.

References

Hageman, Ir. B. BP. 1969: Development of the western part of the Netherlands during the Holocene. *Geologie en Mijinbouw*, 48, 373–388.

Hamada, T. 1977: The Holocene corals of raised reefs of Japan. *Proc. 2nd Int'l. Symp. Fossil Coral and Reefs*. No. 89, 389–395.

Hamano, Y., Maeda, Y., Matsumoto, E. and Kumano, S. 1985: Holocene sedimentary history of some coastal plains in Hokkaido, Japan. III. Transition of diatom assemblages in Tokoro along the Okhotsk Sea. *Japan Jour. Ecol.*, 35, 307–316.

Haraguchi, T. and Iwamatsu, A. 2011: *Detailed maps of the impacts of the 2011 Japan Tsunami Vol. 1: Aomori, Iwate and Miyagi prefectures ; Vol. 2 : Fukushima, Ibaraki and Chiba prefectures*. (Kokon-Shoin), 167p.; 97p. (J)

Hayashi, M. 1991: Geomorphic development of the Izumo plain, Western Japan. *Geogr. Rev. Japan*, 64A, 26–46. (J+E)

Hesp, P. A. 1988: Surfzone, beach and foredune interactions on the Australian south east coast. *Jour. Coastal Res., Special Issue*, 3, 15–25.

Hesp. P. A. and Short, A. D. 1999: Barrier morphodynamics. In Short, A. D. (ed.) *Handbook of Beach and Shoreface Morphodynamics*. (John Wiley and Sons), 307–333.

Hirai, Y. 1983: Lacustrine and sublacustrine microforms of Lake Ogawara with relation to the lake-level changes after the maximum postglacial transgression. *Annals, Tohoku Geogr. Assoc.*, 35, 82–91. (J+E)

Hirai, Y. 1987: Lacustrine and sublacustrine microforms and deposits near the shoreline of Lake Saroma and the sea level changes in the Sea of Okhotsk in late Holocene. *Annals, Tohoku Geogr. Assoc.*, 39, 1–15. (J+E)

Hirai, Y. 1994: Geomorphological development of the lagoons in Japan. *Mem. Fac. Ehime Univ., Nat. Sci.*, 14, 1–71. (J+E)

Hirai, Y. 1995: *Lake Environment of the Coastal Lagoons in Japan*. (Kokon-Shoin), 186p. (J)

Ikeya, N. 1977: Ecology of foraminifera in the Hamana Lake region on the Pacific coast of Japan. *Rep. Facul. Sci., Shizuoka Univ.*, 11, 131–159.

Ikeya, N. and Handa, T. 1972: Surface sediments in Hamana Lake, Pacific coast of Central Japan. *Rep. Facul. Sci., Shizuoka Univ.*, 7, 129–148.

Ikeya, N., Wada, H., Akutsu, H. and Takahashi, M. 1990: Origin and sedimentary history of Hamana-ko Bay, Pacific coast of central Japan. *Mem. Geol. Soc. Japan*, 36, 129–150. (J+E)

Institute of Palaeoenvironmental Research 1994: Analyses of the deposits of the

Kajiko Site. In Hamamatsu City Museum (ed.), *Kajiko Site IX*, 117–134.

Ishii, Y., Ito, A., Nakanishi, T., Hong, W. and Hori, K. 2014: Changes in sedimentary environment and sedimentation rate of the IK1 core obtained from the inner part of Ishikari lowland, northern Japan. *The Quat. Res.*, 53, 143–156. (J+E)

Ishizuka, T. and Kashima, K. 1986: Holocene water level changes of Lake Ogawara presumed by diatom analysis, eastern part of Aomori Prefecture, North Japan. *Geogr. Rev., Japan*, 59, 205–212. (J+E)

Ito, A. 1999: Geomorphic development of the Kitakami River Lowland and Holocene relative sea-level change around Sendai Bay, Northeastern Japan. *Sci. Rep., Tohoku Univ., 7th Series (Geography)*, 49, 81–98.

Ito, A. 2003: Periods of beach ridge range formation on the Kitakami River lowlands and late Holocene sea-level changes around Sendai Bay, Northeastern Japan. *Geogr. Rev. Japan*, 76, 537–550. (J+E)

Japan Meteorological Agency (JMA) 2012: Report on tsunami height from The 2011 off the Pacific coast of Tohoku Earthquake.
www.data.jma.go.jp/svd/eqev/data/2011_3_11_tohoku/index.html

Jelgersma, S. and Van Regteren, A. J. F. 1969: An outline of the geological history of the coastal dunes in the western Netherlands. *Geologie en Mijinbouw*, 48, 335–342.

Kaizuka, S. 2011: *Natural History of Tokyo*. (Kodansha), 327p. (J)

Kaizuka, S. and Moriyama, A. 1969: Geomorphology and subsurface geology of the alluvial plain of the lower Sagami River, Central Japan. *Geogr. Rev., Japan*, 42, 85–105. (J+E)

Kaizuka, S., Naruse, Y. and Matsuda, I. 1977: Recent formations and their basal topography in and around Tokyo Bay, Central Japan. *Quat. Res.*, 8, 32–50.

Kaizuka, S., Akutsu, J., Sugihara, S. and Moriwaki, H. 1979: Geomorphic development of alluvial plains and coasts during the Holocene in Chiba Prefecture, Central Japan, with a note on diatom assemblages in Holocene deposits near the junction of rivers Miyako and Furuyama. *The Quat. Res.*, 17, 189–205. (J+E)

Kajiyama, H. and Itihara, M. 1972: The development history of the Osaka plain with references to the radio-carbon dates. *Mem. Geol. Soc. Japan*, 7, 101–112. (J+E)

Kasai, M. and Koiwa, N. 2014: Holocene geomorphological evolution in the downstream part of the Iwaki Lowland in northeastern Japan. *The Quat. Res.*, 53, 213–228. (J+E)

REFERENCES

Kaseno, Y., Kojima, K., Nakagawa, K. and Miyata, T. 1990: Lagoon Kahoku-gata, Ishikawa Prefecture—Geologic history, geotechnics and the aqueous environment after reclamation works. *Mem. Geol. Soc. Japan*, 36, 35–45. (J+E)

Kayane, H. 1991: Change in molluscan assemblages in the Holocene accompanied by the formation of Futtu Spit, Boso Peninsula. *The Quat. Res.*, 30, 265–280. (J+E)

Kayane, H. and Yoshikawa, T. 1986: Comparative study between present and emergent erosional landforms on the southeast coast of Boso Peninsula, central Japan, *Geogr. Rev., Japan,* 59, 18–36. (J+E)

Kikuchi, M. 2001: Settlement location at Boso Peninsula in the Jomon period: Tokyo Bay and Kujukuri Plain of the coastal Simosa Upland. *The Quat. Res.*, 40, 171–183. (J+E)

Kobayashi, M., Hatano, S., Ichikawa, S. and Kumaki, Y. 1982: Morphotectonic investigation along the coast in the Kanto and Tokai regions. In The Science and Technology Agency (ed.), *Research of Seismotectonics in the Northern Margin of the Philippine Sea Plate.*, 117–131. (J)

Kobayashi, I., Kanzo, K., Kamoi, Y. and Watanabe, Y. 1993: Natural environment and history of Lake Kamo, Sado Island. *Mem. Geol. Soc. Japan*, 39, 89–102. (J+E)

Koike, K. and Ota, Y. ed. 1996: *Coastal Changes in Japan.* (Kokon-Shoin), 185p. (J)

Koiwa, N., Kasai, M. and Ito, A. 2014: Stratified lacustrine water structure and geomorphic environments of the Holocene Lake Jusanko, northeastern Japan. *The Quat. Res.*, 53, 21–34. (J+E)

Komatsubara, J., Shishikura, M. and Okamura, Y. 2007: Activity of Fujikawa-kako fault zone inferred from submergence history of Ukishima-ga-hara lowland, central Japan. *Annual Report on Active Fault and Paleoearthquake Researches No. 7* (AIST Geological Survey of Japan), 119–128.(J+E)

Maeda, Y., Yamashita, K., Matsushima, Y. and Watanabe, M. 1983: Marine transgression over Mazukari shell mound on the Chita Peninsula, Aichi Prefecture, Central Japan. *The Quat.Res.*, 22, 213–222. (J+E)

Masai, Y. 2003: *Geography of Tokyo.* (Seishun-shuppansha), 95p. (J)

Mason, O. K. 1993: The geoarchaeology of beach ridges and cheniers: Studies of coastal evolution using archaeological data. *Jour. Coast. Res.*, 9 , 126–146.

Mason, O. K. and Jordan, J. W. 1993: Heightened North Pacific storminess during synchronous late Holocene erosion of Northwest Alaska beach ridges. *Quat. Res.*, 4, 55–69.

Masuda, F., Fujiwara, O., Sakai, T., Araya, T., Tamura, T. and Kamataki, T. 2001a: Progradation of the Holocene beach-shoreface system in the Kujukuri strand plain, Pacific coast of the Boso Peninsula, central Japan. *The Quat. Res.*, 40, 223–233. (J+E)

Masuda, F., Fujiwara, O., Sakai, T. and Araya, T. 2001b: Relative sea-level changes and co-seismic uplifts over six millennia, preserved in beach deposits of the Kujukuri strand plain, Pacific coast of the Boso Peninsula, Japan. *Jour. Geogr.*, 110, 650–664. (J+E)

Masuda, F., Nakagawa, Y., Sakamoto, T., Ito, Y., Sakurai, M. and Mitamura, M. 2013: Tenma spit deposits in Holocene of the Osaka Plain: Distribution and stratigraphy. *Jour. Sed. Soc. Japan*, 72, 113–123. (J+E)

Matoba, Y. 1970: Distribution of recent shallow water foraminifera of Matsushima Bay, Miyagi Prefecture, Northeast Japan. *Tohoku Univ., Sci. Rep. 2nd ser. (Geol.)*, 42, 1–85.

Matsubara, A. 1984: Geomorphic development of the alluvial plain facing the inner Suruga Bay. *Geogr. Rev., Japan*, 57, 37–56. (J+E)

Matsubara, A. 1988: Geomorphic development of barriers in the coastal lowlands during the Holocene—A case study of the coastal lowlands along the Suruga Bay, Central Japan—*Bull. Dep. Geogr. Univ. Tokyo*, 20, 57–77.

Mastubara, A. 1992: Environmental change and human activity around Megazuka in the Ukishimagahara lowland, Shizuoka Prefecture. *The Quat. Res.*, 31, 221–227. (J+E)

Matusbara, A. 1998: Palaeoenvironmet around the archaeological sites in Kano river delta—A case study of Sanmaibashi Castle—*Hiyoshi Rev. Soc. Sci., Keio Univ.*, 8, 22–33. (J)

Matsubara, A. 1999: Palaeogeographical changes in the Seishin lowland along the Suruga Bay. *Hiyoshi Rev. Soc. Sci., Keio Univ.*, 9, 1–19. (J)

Matsubara, A. 2000a: Holocene geomorphic development of coastal barriers in Japan. *Geogr. Rev., Japan*, 73, 409–434. (J+E)

Matsubara, A. 2000b: Relationships between geomorphic development and human activities—Case studies of coastal lowlands along Suruga Bay—*Hiyoshi Rev. Soc. Sci., Keio Univ.*, 10, 25–40. (J)

Matsubara, A. 2001: Coastal barriers in Hamana Lake and the Hamamatsu lowland. *Hiyoshi Rev. Soc. Sci., Keio Univ.*, 11, 20–32. (J)

Matsubara, A. 2002: Holocene geomorphic development of coastal barriers in

Japan. *Hiyoshi Rev. Soc. Sci., Keio Univ.*, 12, 37–68.

Matsubara, A. 2003: Relationships between Holocene geomorphic development of coastal ridges and human activities—A case study of the coastal lowlands along Suruga Bay, Central Japan—*Hiyoshi Rev. Soc. Sci., Keio Univ.*, 13, 23–40.

Matsubara, A. 2004: Sedimentary environment around archaeological sites in the Hamamatsu lowland. *Hiyoshi Rev. Soc. Sci., Keio Univ.*, 14, 35–52. (J)

Matsubara, A. 2005: Processes in the Holocene development of coastal ridges in Japan. *Hiyoshi Rev. Soc. Sci., Keio Univ.*, 15, 73–90.

Matsubara, A. 2008: Relationships between geomorphic development of coastal ridges and human activities—A case study of the Hamamatsu and Haibara Lowlands—*Hiyoshi Rev. Soc. Sci., Keio Univ.*, 18, 1–13. (J)

Matsubara, A. 2009a: Geomorphic development of coastal ridges along the Mano Bay, Sado Island. *Hiyoshi Rev. Soc. Sci., Keio Univ.*, 19, 1–13. (J)

Mastubara, A. 2009b: Fossil foraminiferal assemblages in Kasumigaura Lowland. In Miho Village Board of Education (ed.), *Okadaira Shell Mound*. (Miho Village), 23–32. (J)

Matsubara, A. 2012: Landforms of the plain along the west part of Tokyo Bay—Case studies of the east part of Tokyo and the Tsurumi River Basin—*Hiyoshi Rev. Soc. Sci., Keio Univ.*, 22, 1–12. (J)

Matsubara, A. 2013: Holocene geomorphic development of beach ridge plains along the Tokyo Bay—Case studies of the Tateyama and Obitsu River Lowlands—*Hiyoshi Rev. Soc. Sci., Keio Univ.*, 23, 1–14. (J)

Matsubara, A. 2014: Landforms and distribution of archaeological sites in the Kujukuri coastal lowland. *Hiyoshi Rev. Soc. Sci., Keio Univ.*, 24, 21–28. (J)

Matsubara, A., Matsushima, Y., Ishibashi, K., Moriwaki, H. and Kashima, K. 1986: Environmental changes during the Holocene in the Matsuzaki Lowland, western part of the Izu Peninsula, Central Japan. *Jour. Geogr.*, 95, 339–356. (J+E)

Matsubara, H.and Shiomi, R. 2010: Fossil molluscan assemblages of the Iwataki bore hole core and palaeoenvironment of the Aso-Kai. In Uemura, Y. (ed.), *Palaeoenvironment and Geomorphic Development of the Kumihama Bay—in Comparison with the Aso-Kai and Amanohashidate*. (Kyotango City Board of Education), 105–110. (J)

Matsuda, I. 2009: *Geomorphology of Edo-Tokyo*. (Korejio), 318p. (J)

Matsuda, T., Yui, M., Matsushima, Y., Imanaga, I., Hirata, D., Togo, M., Kashima, K., Matsubara, A., Nakai, N., Nakamura, T. and Matsuoka, K. 1988: Subsurface study of Isehara Fault, Kanagawa Prefecture, detected by drilling—Depositional environments during the last 7000 years and fault displacement associated with the Gangyo Earthquake in A.D. 878—*Bull. Earthq. Res. Inst., Univ. Tokyo*, 63, 145-182. (J+E)

Matsumoto, H. 1981: Sea-level changes during the Holocene and geomorphic development of the Sendai coastal plain, Northeast Japan. *Geogr. Rev. Japan*, 54, 72-85. (J+E)

Matsumoto, H. 1984: Beach ridge ranges on Holocene coastal plains in northeast Japan—The formative factors and periods—*Geogr. Rev., Japan*, 57, 720-738. (J+E)

Matsunaga, K and Ota, Y. 2001: Late Holocene environmental changes at Kuninaka Plain, Sado Island, Central Japan, deduced from sediment facies and diatom assemblages. *The Quat. Res.*, 40, 355-371. (J+E)

Matsushima, Y. 1984a: Neotectonics deduced from Holocene marine terraces along the coast of Sagami Bay, Central Japan. *The Quat. Res.*, 23, 165-174. (J+E)

Matsushima, Y. 1984b: Shallow marine molluscan assemblages of Postglacial period in the Japanese Islands—Its historical and geographical changes induced by the environmental changes—*Bull. Kanagawa Pref. Museum (Natural Science)*, No. 15, 37-109. (J+E)

Matsushima, Y. and Yoshimura, M. 1979: Radiocarbon ages of the Numa formation along the Heguri River, Tateyama, Chiba Prefecture. *Bull. Kanagawa Pref. Museum (Natural Science)*, No. 11, 1-9. (J+E)

Matsushima, Y. and Kitazato, H. 1980: Environmental changes in Utsumi. In Minami-Chita Town Board of Education (ed.), *Mazukari Shell Mound*, 113-114. (J)

Matsushita, K. 1979: Buried landforms and the upper Pleistocene-Holocene landforms deposits of the Ishikari coastal plain, Hokkaido, North Japan. *The Quat. Res.*, 18, 69-78. (J+E)

Mii, H. 1960: Holocene deposits of Hachiro-gata (Lagoon). *Sci. Rep., Tohoku Univ., 2nd Ser (Geol.) Spec.* 4, 590-598. (J+E)

Minoura, K. and Nakaya, S. 1990: Origin of inter-tidal lake and marsh environments and around Lake Jusan, Tsugaru. *Mem. Geol. Soc. Japan*, 36, 71-87. (J+E)

Mizuno, A., Ohshima, K., Nakano, S., Noguchi, Y., and Masaoka, E. 1972: Late Quaternary history of the brackish lakes Naka-umi and Shinji-ko on San'in coastal plain and related some problems. *Mem. Geol. Soc. Japan*, 7, 113-124. (J+E)

Moriwaki, H. 1979: The landform evolution of the Kujukuri coastal plain, Central Japan. *The Quat. Res.*, 18, 1–16. (J+E)

Moriwaki, H. 1982: Geomorphic development of Holocene coastal plains in Japan. *Geogr. Rep. Tokyo Metropolitan Univ.*, 17, 1–42.

Morton, R. A. 1994: Texas barriers. In Davis, R. A. (ed.), *Geology of Holocene Barrier Island Systems.* (Springer-Verlag), 74–114.

Moslow, T. F. and Colquhoun, D. J. 1981: Influence of sea level change on barrier island evolution. *Oceanis*, 7, 439–454.

Moslow, T. F. and Herson, S. D. 1994: The outer banks of North Carolina. In Davis, R. A. (ed.), *Geology of Holocene Barrier Island Systems.* (Springer-Verlag), 47–74.

Nagaoka, S., Maemoku, H. and Matsushima, Y. 1991: Evolution of Holocene coastal landforms in the Miyazaki plain, Southern Japan. *The Quat. Res.*, 30, 59–78. (J+E)

Nagaoka, S., Yokoyama, Y., Nakada, M., Maeda, Y., Okuno, J. and Shirai, K. 1997: Holocene geomorphic development and sea-level change in the Tamana plain, southeastern coast of Ariake Bay, Western Japan. *Geogr. Rev. Japan*, 70, 287–306. (J+E)

Nagasako, T., Okuno, M., Moriwaki, H., Arai, F. and Nakamura, T. 1999: Paleogeography and tephras of the Kimotsuki lowland, southern Kyushu, Japan, in the middle to late Holocene. *The Quat. Res.*, 38, 163–173. (J+E)

Nakagawa, T. 1987: Late Pleistocene and Holocene developments of Niigata Plain, Central Japan. *Jour. Geol. Soc. Japan*, 93, 575–586. (J+E)

Nakata, T., Koba, M., Imaizumi, T., Jo. W. R., Matsumoto, H. and Suganuma, T. 1980: Holocene marine terraces and seismic crustal movements in the southern part of Boso Peninsula, Kanto, Japan. *Geogr. Rev., Japan*, 53, 29–44. (J+E)

Nguyen, V. L., Tateishi, M. and Kobayashi, I. 1998: Reconstruction of sedimentary environments for late Pleistocene to Holocene coastal deposits of Lake Kamo, Sado Island, Central Japan. *The Quat. Res.*, 37, 77–94.

Niigata Ancient Dune Research Group 1974: Niigata sand dunes and archaeological relics—The geohistory of the formation of Niigata sand dunes, Part I—. *The Quat. Res.*, 13, 57–65. (J+E)

Ohira, A. 1992: Geomorphic development of the northeastern part of the Niigata plain, Central Japan, during the Holocene. *Geogr. Rev., Japan*, 65, 876–888. (J+E)

Ohira, A. 1995: Holocene evolution of peatland and paleoenvironmental changes in the Sarobetsu lowland Hokkaido, Northern Japan. *Geogr. Rev., Japan*, 68, 695-712. (J+E)

Ohira, A., Umitsu, M. and Hamade, S. 1994: Late Holocene development of peatlands in small alluvial lowlands around Lake Furen, eastern Hokkaido, Japan. *The Quat. Res.*, 33, 45-50. (J)

Okazaki, Y. 1960a: Topographic development of the Kushiro moor and its surroundings, Hokkaido, Japan. *Geogr. Rev., Japan*, 33, 462-473. (J+E)

Okazaki, Y. 1960b: Paleogeography of the Kushiro plain, Hokkaido in the alluvial age. *The Quat. Res.*, 1, 255-262. (J+E)

Omoto, K. 1976: Tohoku University radiocarbon measurements III. *Sci. Rep. Tohoku Univ., Ser. 7 (Geography)*, 26, 135-157.

Onuki, Y., Mii, H., Shimada, I., Takeuti, S., Ishida, T. and Saito, T. 1963: Late Quaternary history of Tsugaru Jusanko (Jusan Lagoon) district, Aomori Prefecture, Japan. *Tohoku Univ., Sci. Rep. (Geol.)*, 58, 1-85. (J+E)

Ota, Y. and Seto, N. 1968: Note on the problems of sand dunes at the coastal area along the Sagami Bay, Kanagawa Prefecture, Central Japan. *Bull. Yokohama National Univ. Sci., Sec. II.*, 14, 35-60. (J+E)

Ota, Y., Matsushima, Y. and Moriwaki, H. eds. 1981: *Atlas of Holocene Sea Level Records in Japan*. Working Group of Project 61, Holocene Sea Level Project, IGCP, 195p.

Ota, Y., Matsushima, Y. and Moriwaki, H. 1982: Note on the Holocene sea-level study in Japan—On the basis of "Atlas of Holocene Sea-level Records in Japan" —*The Quat. Res.*, 21, 133-143. (J+E)

Ota, Y., Umitsu, M. and Matsushima, Y. 1990: Recent Japanese research on relative sea level changes in the Holocene and related problems—Review of studies between 1980 and 1988—*The Quat. Res.*, 29, 31-48. (J+E)

Ota, Y., Matsubara, A., Matsushima, Y., Kashima, K., Kanauchi, A., Suzuki, Y., Watanabe, M., Sawa, H. and Azuma, T. 2008: Holocene paleoenvironmental changes at coastal Kuninaka Plain, Sado Island, off central Japan, as deduced from analysis of drilling data. *The Quat. Res.*, 47, 143-157. (J+E)

Pierce, J. W. and Colquhoun, D. J. 1970: Holocene evolution of a portion of the North Carolina coast. *Geol. Soc. Amer. Bull.*, 81, 3697-3714.

Pirazolli, P. A. 1991: *World Atlas of Holocene Sea-level Changes*. (Elsevier), 300p.

Pirazolli, P. A. 1996: *Sea-Level Changes—The Last 20,000 Years*. (John Willey &

Sons), 211p.

Reimer, P. J., Baillie, M. G. L., Bard, E., Bayliss, A., Beck, J. W., Blackwell, P. G., Bronk Ramsey, C., Buck, C. E., Burr, G. S., Edwards, R. L., Friedrich, M., Grootes, P. M., Guilderson, T. P., Hajdas, I., Heaton, T. J., Hogg, A. G., Hughen, K. A., Kaiser, K. F., Kromer, B., McCormac, F. G., Manning, S. W., Reimer, R. W., Richards, D. A., Southon, J. R., Talamo, S., Turney, C. S. M., van der Plicht, J. and Weyhenmeyer, C. E. 2009: IntCal09 and Marine09 Radiocarbon Age Calibration Curves, 0–50,000 Years cal BP. *Radiocarbon*, 51, 1111–1150.

Sadakata, N. 1991: The influence of iron sand mining (Kannna-nagashi) on the formation of the Sotohama beach ridges in the Yumigahama Peninsula of South-western Japan. *Geogr. Rev., Japan*, 64, 759–778

Sadakata, N., Shiragami, H. and Kashima, K. 1988: Sea level of the two stages in the late Holocene deduced from the deposits of the coastal lowland in the subsided region of the southern Shikoku Island. *The Quat. Res.*, 27, 125–129. (J)

Sagayama, T., Tonozaki, T., Kondo, T., Okamura, S. and Sato, K. 2010: Stratigraphy and paleoenvironment of Upper Pleistocene to Holocene sediments in the Ishikari Plain. *Jour. Geol. Soc. Japan*, 116, 13–26. (J+E)

Saito, Y. 1987: Processes on the deposition of marine sediment related to the Holocene sea level changes. *Monthly Chikyu (Earth)*, 9, 533–541. (J)

Saito, Y. 1988: Barrier system as a recorder of Holocene sea-level changes and problems on reconstruction of their paleogeography in the early Holocene: Example from Lake Hamana, Central Japan. *Clastic Sediments (Jour. Res. Gr. Clas. Sed. Japan)*, 5, 109–132. (J+E)

Saito, Y. 1995: High-resolution sequence stratigraphy of an incised-valley fill in a wave-and fluvial-dominated setting: latest Pleistocene-Holocene examples from the Kanto Plain, central Japan. *Mem. Geol. Soc. Japan,* 45, 76–100.

Saito, Y., Inouchi, Y. and Yokota, S. 1990: Coastal lagoon evolution influenced by Holocene sea-level changes, Lake Kasumigaura, Central Japan. *Mem. Geol. Soc. Japan*, 36, 103–118. (J+E)

Sakaguchi, Y. 1961: Paleogeographical studies of peat bogs in northern Japan. *Jour. Fac. Sci., Univ. Tokyo, ser. II*, No. 3, 421–513.

Sakaguchi, Y. 1974: *Paleogeographical of Peat Bogs.* (Tokyo Univ. Press.), 329p. (J)

Sakaguchi, Y., Kashima, K. and Matsubara, A. 1985: Holocene marine deposits in Hokkaido and their sedimentary environments. *Bull. Dept. Geogr. Univ. Tokyo*, 17, 1–17.

Sakaguchi, Y., Kashima, K. and Matsubara, A. 2009: Environmental changes around Okadaira Site. In Miho Village Board of Education (ed.), *Okadaira Shell Mound*. 7–21. (J)

Sanderson, P. G., Eliot, I. and Fuller, M. 1998: Historical development of a foredune plain at Desperate Bay, Western Australia. *Jour. Coastal Res.*, 14, 1187–1201.

Sato, Y., Fujiwara, O., Ono, E. and Umitsu, M. 2011: Environmental changes in coastal lowlands around the Lake Hamana during the middle to late Holocene. *Geogr. Rev., Japan*, 84, 258–273. (J+E)

Sato, Y. and Ono, E. 2013: Late Holocene geoenvironmental changes around Lake Koyama • Tottori Plain, western Japan. *Geogr. Rev., Japan*, 86, 270–287. (J+E)

Sawai, Y., Shishikura, M., Okamura, Y., Takada, K., Matsuura, T., Aung, T., Komatsubara, J., Fujii, Y., Fujiwara, O., Satake, K., Kamataki, T. and Sato, N. 2007: A study on paleotsunami using handy geoslicer in Sendai Plain (Sendai, Natori, Iwanuma, Watari and Yamamoto), Miyagi, Japan. *Annual Report on Active Fault and Paleoearthquake Researches No. 7* (AIST Geological Survey of Japan), 47–80. (J+E)

Sherman, D. J. (ed.) 2013: *Treatise on Geomorphology Vol.10 Coastal Geomorphology* (Academic Press), 448p.

Shibayama, T. and Kayane, H. (ed.), 2013: *Japanese Coasts*. (Asakura-shoten), 152 p. (J+E)

Shiragami, H. 1983: Beach ridges and their regional difference in the northern coast of the Suo-Nada, Seto Inland Sea. *Chiri-Kagaku (Geogr. Sci.)*, 38, 133–141. (J+E)

Shiraishi, T. 1990: Holocene geologic development of the Hachiro-gata lagoon, Akita Prefecture, northeast Honshu, Japan. *Mem. Geol. Soc. Japan*, 36, 47–69. (J+E)

Shishikura, M. and Miyauchi, T. 2001: Holocene geomorphic development related to seismotectonics in coastal lowlands of the Boso Peninsula, Central Japan. *The Quat. Res.*, 40, 235–242. (J+E)

Shishikura, M., Fujiwara, O., Sawai, Y., Namegaya, Y. and Tanigawa, K. 2012: Inland-limit of the tsunami deposit associated with the 2011 Off-Tohoku Earthquake in the Sendai and Ishinomaki Plains, Northeastern Japan. *Annual Report on Active Fault and Paleoearthquake Researches No. 12* (AIST Geological Survey of Japan), 45–61. (J+E)

References

Short, A. D. 2010: Sediment transport around Australia—sources, mechanisms, rates and barrier forms. *Jour. Coastal Res.*, 26, 395–402.

Sugimura, A. and Naruse, Y. 1954, 1955: Changes in sea level, seismic upheavals, and coastal terraces in the southern Kanto region, Japan (I), (II). *Japan. Jour. Geol. Geogr.*, 24, 101–113; 26, 165–176.

Suzuki, T. and Saito, Y. 1987: Heavy mineral composition and provenance of Holocene marine sediments in Lake Kasumigaura, Ibaraki, Japan. *Bull. Geol. Surv. Japan*, 38, 139–164. (J+E)

Takahashi, M. 1982: The geomorphological structure of the Mihara alluvial plain on Awaji Island, Hyogo Prefecture. *Annals, Tohoku Geogr. Assoc.*, 34, 138–150. (J+E)

Takano, S. 1978: Development of the compound spit of Notsukezaki, Hokkaido. *Annals, Tohoku Geogr. Assoc.*, 30, 82–90. (J+E)

Tamura, T., Masuda, F., Sakai, T. and Fujiwara, O. 2003: Temporal development of prograding beach-shoreface deposits: the Holocene of Kujukuri coastal plain, eastern Japan. *Mar. Geol.*, 198, 191–207.

Tamura, T. and Masuda, F. 2005: Bed thickness characteristics of inner-shelf storm deposits associated with a transgressive to regressive Holocene wave-dominated shelf, Sendai coastal plain, Japan. *Sedimentology*, 52, 1375–1395.

Tamura, T., Saito, Y. and Masuda, F. 2006: Stratigraphy and sedimentology of marine deposits in a Holocene strand plain: examples from the Sendai and Kujukurihama coastal plains. *Mem. Geol. Soc. Japan*, 59, 83–92. (J+E)

Tamura, T., Murakami, F., Nanayama, F., Watanabe, K. and Saito, Y. 2008: Ground-penetrating radar profiles of Holocene raised-beach deposits in the Kujukuri strand plain, Pacific coast of eastern Japan. *Mar. Geol.*, 248, 11–27.

Tanabe, S. 2013: Strata formation in a tectonically subsiding coastal lowland: example from alluvium in the Echigo Plain, central Japan. *Jour. Geogr.*, 122, 291–307. (J+E)

Tanaka, H., Hasegawa, T., Kimura, S., Okamoto, I. and Sakai, Y. 1996: The geohistory of the formation of the Niigata sand dunes. *The Quat. Res.*, 35, 207–218. (J+E)

Tanner, W. F. 1988: Beach ridge data and sea level history from the Americas. *Jour. Coast. Res.*, 4, 81–91.

Taylor, M. and Stone, G. W. 1996: Beach ridges: a review. *Jour. Coastal Res.*, 12, 612–621.

Thom, B. G. 1983: Transgressive and regressive stratigraphies of coastal sand barriers in southeast Australia. *Mar. Geol.*, 56, 137–158.

Thom, B. G., Bowman, G. M. and Roy, P. S. 1981: Late Quaternary evolution of coastal sand barriers, Port Stephens: Myall Lakes area, central New South Wales, Australia. *Quat. Res.* 30, 345–364.

Thom, B. G. and Roy, P. S. 1985: Relative sea levels and coastal sedimentation in southeast Australia in the Holocene. *Jour. Sedimentary Petrology*, 55, 257–264.

Tokuoka, T., Onishi, I., Takayasu, K. and Mitsunashi, T. 1990: Natural history and environmental changes of Lake Nakaumi and Shinji. *Mem. Geol. Soc. Japan*, 36, 15–34. (J+E)

Trenhaile, A. S. 1997: *Coastal Dynamics and Landforms*. (Clarendon Press). 366p.

Uda, T. 2010: *Japan's Beach Erosion—Reality and Future Measure—*. Advanced Series on Ocean Engineering, Vol. 31, (World Scientific), 418p.

Uemura, Y. 2010: Outlines of the research of boring around the Aso-Kai. In Uemura, Y. (ed.) *Palaeoenvironment and Geomorphic Development of the Kumihama Bay —in Comparison with the Aso-Kai and Amanohashidate*. (Kyotango City Board of Education), 97–104. (J)

Uesugi, Y. and Endo, K. 1973: On the topographies and soils of the Ishikari coastal plain. *The Quat. Res.*, 12, 115–124. (J+E)

Umitsu, M. 1991: Holocene sea-level changes and coastal evolution in Japan. *The Quat. Res.*, 30, 187–196.

Urabe, A. 2008: Holocene depositional systems along the Agano River of the Echigo Plain, central Japan. *The Quat. Res.*, 47, 191–201. (J+E)

Urabe, A., Yoshida, M. and Takahama, N. 2006: Development process of barrier-lagoon system in the Holocene sediments of the Echigo Plain, central Japan. *Mem. Geol. Soc. Japan* , 59, 111–127. (J+E)

Urabe, A., Fujimoto, Y. and Kataoka, K. 2011: Influence of a volcanogenic flood event on an alluvial depositional system: the Holocene Echigo Plain of northeast Japan. *Jour. Geol. Soc. Japan*, 117, 483–494. (J+E)

Van Straaten, L. M. J. U. 1965: Coastal barrier deposits in south and north Holland. *Meded. Van Geologische Stichting*, 17, 41–45.

Wilkinson, B. H. 1975: Matagorda Island, Texas: the evolution of a Gulf Coast barrier complex. *Geol. Soc. of Amer. Bull.*, 86, 959–967.

Yasui, K., Kamoi, Y., Kobayashi, I., Urabe, A., Watanabe, H. and Mikata, I. 2002: Formation and environmental changes of a Holocene brackish lake in the northern Echigo Plain, central Japan. *The Quat. Res.*, 41, 185–197. (J+E)

Yokota, K. 1978: Holocene coastal terraces on the southeast coast of the Boso Peninsula. *Geogr. Rev., Japan*, 51, 349–364. (J+E)

Yonekura, N. 1975: Quaternary tectonic movements in the outer area of southwest Japan with special reference to seismic crustal deformation. *Bull. Dept. Geogr., Univ. Tokyo*, 7, 19–71.

Yonekura, N., Ikeda, Y., Kashima, K. and Matsubara, A. 1985: Boring of alluvial deposits in the coastal lowlands around Suruga Bay, Central Japan. In Sakaguchi, Y. (ed.), *Environmental Changes since the Last Glacial Period*. (Univ. Tokyo), 35–80. (J)

Yoshida, M., Hoyanagi, K., Urabe, A., Yamazaki, A., Yamagishi, M. and Omura, A. 2006: Reconstruction of sedimentary environments on the basis of sedimentary facies, TOC, TN and TS contents; examples from the latest Pleistocene and Holocene sediments in the Niigata Plain, central Japan. *Mem. Geol. Soc. Japan*, 59, 93–109. (J+E)

Yoshikawa, T., Kaizuka, S. and Ota, Y. 1981: *The Landforms of Japan*. (Univ. Tokyo Press), 222p.

Note: (J): in Japanese, (J+E): in Japanese with English abstract.

Index

A
Abe River, 10, 53, 126
Abe River Fan, 127
abrasion platform, 19, 45, 51, 53, 86, 94, 97, 117, 121, 147, 149
Abukuma River, 51, 132
Amanohashidate, 7, 27, 129
Ara River, 114
Ashitaka Mountain, 29, 105
Aso-Kai (Aso Lake), 7, 27, 130

B
beach nourishment, 125, 129
Boso Peninsula, 65, 97, 115
breakwater, 125, 129

C
coastal dike, 129
coastal erosion, 125, 126, 129

E
Edo-Maejima (Edo Island), 117
Edo period, 121
Eurasian Plate, 11, 123

F
Foraminifera, 12
Fossa Magna, 6, 123
Fuji Mountain, 10, 29, 32, 106, 114
Fuji River, 10, 32, 123

G
Genroku Kanto Earthquake, 68, 112

H
Haibara Lowland, 7, 10, 84, 112, 146, 149
Hakone Mountain, 29, 114
Hamamatsu Lowland, 7, 19, 101, 147
Hamana Lake, 7, 17, 101, 147
Heguri River, 65, 69
Hibiya-Irie (Hibiya Inlet), 117, 121
Higashimae Site, 103
Hirose River, 51, 132
Holocene Transgression (Jomon Transgression), 3, 97, 116, 117, 126, 134, 139, 141, 143, 144
Hongo Upland, 117

I
Iba Site, 103
Imperial Palace (Edo Castle), 117, 123
Isarago Shell Mound, 121
Ishikawa II Archaeological Site, 110
Ishinomaki Lowland, 133
Izu Peninsula, 10, 29
Izu-Ogasawara Trench, 6

J
Japan Trench, 6, 132
Jogan Earthquake, 133
Jomon Period, 103, 104, 106, 110, 112, 114

K
Kajiko Archaeological Site, 102
Kajiko-kita Archaeological Site, 102

Kamezuka Archaeological Site, 122
Kamo Lake, 44
Kanda River, 117
Kano River, 10, 29, 123
Kano River Delta, 32
Kano River Lowland, 7, 10, 29, 104, 148
Kanto Basin, 73
Kanto Loam, 114
Kanto Plain, 73, 114
Katsumata River, 86
Kawagodaira Pumice (Kg), 29, 37, 105
Kise River, 10, 29
Kofun Period, 105, 106, 110, 112
Kokufu River, 44
Kujukurihama Lowland, 7, 73, 114, 134
Kuninaka Lowland, 7, 44, 114

L
Last Glacial Maximum, 116
Last Glacial Stage, 1
Last Interglacial Stage, 116

M
Mano Bay, 44
Matsuzaki Lowland, 7, 10, 77, 146
Megazuka Archaeological Site, 105, 148
Meiji Era, 123
Metropolis of Tokyo, 116
Miho Spit, 11, 53, 110, 126, 148
Mishima Fan, 32
Mishima Lava, 32
Miura Peninsula, 69, 115
Miyamichi Archaeological Site, 110
Miyazu Bay, 27, 129, 130
Musashino Terrace, 116

N
Naka River, 77

Nanakita River, 51, 132
Natori River, 51, 132
North American Plate, 123, 132
Numa Fossil Reef Coral, 69
Numa Marine Terrace, 73, 112
Numazu Castle, 104
Numazu-Fuji Coast, 123

O
Obaradai Terrace, 116
Obitsu River, 97, 115
Obitsu River Delta, 97
Obitsu River Lowland, 7, 97, 114
Obuchi Scoria (ObS), 32, 37, 106
Oi River, 10, 86, 147
Old Kitakami River, 134

P
Pacific Plate, 132
Palaeo-Tokyo Bay, 116
Philippine Sea Plate, 11, 123
Post-glacial Stage, 116

R
Raitokoro River, 41
relative sea level curve, 139
Ryotsu Bay, 44

S
Sagami Bay, 92, 114
Sagami River, 94, 114
Sagami River Lowland, 7, 92
Sagami Trough, 6, 65
sand bypass, 129
Sanmaibashi Castle, 104
Saroma Lake, 41
sea level change, 2, 139
Sendai Bay, 51, 132
Sendai Lowland, 7, 51, 132
shell mound, 121
Shibuya-Furu River, 117
Shimizu Lowland, 7, 10, 52, 109, 126,

139, 148
Shimosueyoshi Terrace, 116
Shinmeizuka Archaeological Site, 106
Shirayuri Archaeological Site, 112
Showa Period, 123, 129
Sumida River, 117, 123
Suruga (-Nankai) Trough, 6, 10, 11, 123, 145, 146
Suruga Bay, 11, 29, 52, 69, 77, 84, 123, 145

T
Tachikawa Terrace, 116
Taisho Kanto Earthquake, 68
Tama River, 114
Tateyama Lowland, 7, 65, 112
tectonic movement, 145
Tennnouzan Archaeological Site, 110
Tohoku Earthquake, 132, 149
Tokoro Lowland, 7, 40
Tokoro River, 40
Tokyo Bay, 65, 97, 114, 116, 117, 121, 123
Tomoe River, 10, 52, 110
Tone River, 114
tsunami, 123, 132, 134, 149

U
Udo Hill, 53, 110
Ukishimagahara Lowland, 7, 10, 29, 104, 145, 146, 148

Y
Yayoi Period, 102, 103, 106, 109, 110, 112

Holocene Geomorphic Development of Coastal Ridges in Japan

2015年3月30日　初版第1刷発行

著　者————松原彰子
発行者————坂上　弘
発行所————慶應義塾大学出版会株式会社
　　　　　　108-8346　東京都港区三田 2-19-30
　　　　　　TEL〔編集部〕03-3451-0931
　　　　　　　　〔営業部〕03-3451-3584〈ご注文〉
　　　　　　　　〔　〃　〕03-3451-6926
　　　　　　FAX〔営業部〕03-3451-3122
　　　　　　振替 00190-8-155497
　　　　　　URL http://www.keio-up.co.jp/
装　丁————耳塚有里
印刷・製本——株式会社加藤文明社
カバー印刷——株式会社太平印刷社.

Ⓒ2015 Akiko Matsubara
Printed in Japan　ISBN 978-4-7664-2215-3